First published 2009 by
PALGRAVE MACMILLAN

Palgrave Macmillan in the UK is an imprint of Macmillan Publishers Limited, registered in England, company number 785998, of Houndmills, Basingstoke, Hampshire RG21 6XS.

Palgrave Macmillan in the US is a division of St Martin's Press LLC, 175 Fifth Avenue, New York, NY 10010.

Palgrave Macmillan is the global academic imprint of the above companies and has companies and representatives throughout the world.

Palgrave® and Macmillan® are registered trademarks in the United States, the United Kingdom, Europe and other countries

ISBN-13: 978-0-230-20223-8   hardback
ISBN-10: 0-230-20223-3   hardback

This book is printed on paper suitable for recycling and made from fully managed and sustained forest sources. Logging, pulping and manufacturing processes are expected to conform to the environmental regulations of the country of origin.

A catalogue record for this book is available from the British Library.

A catalogue record for this book is available from the Library of Congress.

10   9   8   7   6   5   4   3   2   1
18   17   16   15   14   13   12   11   10   09

Printed and bound in Great Britain by
CPI Antony Rowe, Chippenham and Eastbourne

Contents

List of Tables

Preface

Our story of sharefarming in England has taken nearly 30 years to crystallize and almost merits a history in its own right. It is a classic example of a research student stumbling on an idea, hitherto entirely overlooked by historians, which gradually overturns the received wisdom; in this case the absence of sharefarming in England. The idea that sharefarming existed, and was considered sufficiently significant for a Norfolk landowner to copy out lengthy sharefarming agreements into his estate book on the eve of the agricultural revolution, was a radical proposition for the agricultural history establishment to accept. In 1996 Dr Joan Thirsk, who had examined Liz Griffiths' PhD thesis, encouraged her to give a paper to the British Agricultural History Society. It was received politely, but raised few questions and made hardly any converts. Undeterred, Joan introduced Liz to French historians specializing in *métayage*, which led to further papers, the publication of the sharefarming agreements in French, and an understanding of the true importance of the discovery in a European context. Soon after, the Norfolk Record Society agreed to publish William Windham's Green Book, which contained most of the agreements, in its entirety. In 2001, Joan asked Liz to join her in giving a paper at the Economic History Conference on a neglected area of agricultural history. The response was gratifying, as sharefarming had become in the intervening years a recognized practice in farming circles. The real breakthrough came in 2002 when the Marquess Townshend of Raynham kindly allowed access to the Townshend archives. Here was significant evidence of sharefarming from the 1660s to the 1690s. This was enough to convince Mark Overton that we should bid for a research grant to take the project forward. We are very grateful to the ESRC for funding the project, 'Farming to halves: the hidden history of sharefarming in England from medieval to modern times' (RES-000-23-1231) with Mark Overton and Professor Michael Winter as co-investigators, from 2005–7.

This book is therefore the result of a tortuous but tenacious journey. At every stage, we have received encouragement and support from fellow historians. Without Joan Thirsk sharefarming would never have reached the public domain, but others have also played a vital role, allowing access to documents, giving interviews, sending references, reading chapters, and just talking things through. Our particular thanks go to the Marquess Townshend of Raynham, and his secretary, Scilla Landale, for access to the Townshend archive; in Shropshire to the Evans family at Curdale Farm, Cleobury Mortimer, John Griffiths at Overwood Farm, Cleobury Mortimer and David Cooke of Coppice Farm, Ratlinghope for allowing us to use their farm diaries; to Peter Edwards for lending us his notes on Shropshire probate inventories; to John Alban and the staff at the Norfolk Archive Centre; to the staff at the Shropshire, Hereford, Lincolnshire and Sussex Record Offices, and the libraries at Newtown, Powys and Hereford. Our thanks also go to those that gave oral evidence: in Shropshire, John Haywood of Wall Town Farm, Neen Savage; Brian Price of Bockleton Court, Stoke St. Milborough; Brian Davis of Penywern, Clun; and Jane Bevan, sister of Edward Foster of Newton House, Bridgnorth: in Norfolk, to Jim Papworth of Felmingham Hall and the late Ian MacNicol of Stody Lodge. We received helpful information from John Henderson of Skipton, North Yorkshire; John Cyster of Newenden, Kent; John Young, previously head of Land Agency and Agriculture for the National Trust; Peter Fletcher, of Stratton and Holborow; Ian Hamilton, of Hamilton Taylor; Will Gemmill of Strutt and Parker; Philip Wynn of Aubourn, FPD Savills and the Country Land and Business Association.

A number of historians kept a watchful eye out for references and provided useful leads, including John Broad, Richard Hoyle, Paul and Elizabeth Rutledge, Susannah Wade Martins, A. Hassell Smith, Robin Stanes, and Nat Alcock who led us to Richard Suggett's book on Radnorshire. Several European colleagues, including Francesco Galassi, Annie Antoine, Rui Santos, and François Brumont were especially helpful. More thanks go to those who replied to our updates in *Rural History Today* and sent references: Christopher Smout, Avice Wilson, Jim Lewis, Angela Hall, Peter Annels, and Chris Lewis. Our thanks also go to Richard Wilson who went through the whole text at an early stage, and to Robert Kirkham for keeping us up to date with events in New Zealand.

Professor Michael Winter, Director of the Centre for Rural Policy Research at the University of Exeter, was a co-applicant on our ESRC grant, and encouraged us throughout the duration of the ESRC project. We are very grateful to him for contributing the unpublished material from the 2007 survey of farm tenures in Chapter 9, and to Dr Allan Butler for carrying out the analysis of the data.

Finally our thanks to Meemee Overton for her support and to Peter Griffiths who not only provided access to his family archives and technical advice on land management, but has lived with sharefarming for as long as Liz has known him.

1
Introduction

Farming to halves is the English version of sharefarming, a system of letting land which is common in Europe and the New World but thought not to have existed in England until very recently. According to Arthur Young the absence of sharefarming in England was the point of difference between English and French agriculture, and one that explained the success of the former and the failure of the latter. In his *Travels through France* of the late 1780s he famously argued that *métayage*, as sharefarming is known in that country, was the root cause of the backward nature of French farming. 'The poor people who cultivate here are *métayers*, that is men who hire the land without the ability to stock it; the proprietor is forced to provide the cattle and seed and he and his tenant divide the produce ... It is an absurd system which perpetuates poverty and excludes instruction. In this most miserable of all the modes of letting land the defrauded landlord receives a contemptible rent, the farmer is in the lowest state of poverty, the land is miserably cultivated and the nation suffers as severely as the parties themselves.'[1]

As Young made clear to his French hosts, the English were fortunate to have avoided such a deplorable system. In England, landowners leased their farms at fixed money rents for varying terms of years and provided the land and buildings, leaving the responsibility for working capital to the tenant. In this arrangement, fixed and working capital remained entirely separate with benefits to both parties. The landowner retained control over his capital and received a guaranteed income while the tenant bore the risk of farming and profited from his efforts. The system encouraged investment, promoted improvement and underpinned a revolution in agricultural production. In this way English landlords achieved rent levels that were the envy of Europe, 'land which in England would let at 10 shillings, pays about 2s 6d for both land and livestock in France'.[2]

Generations of Frenchmen have shared Young's assessment of French failure and English success. With *métayage*, as he explained, there was no clear distinction between the role of landowner and farmer. Invariably, the landowner provided the fixed assets, contributed to the cost of cultivation, livestock and seed, and divided the crop with his *métayer* at the end of the year. Thus they shared the risk of farming. Although the system was more equitable in theory, in practice it discouraged the parties from investing capital and labour. The landowner depended on the farmer's competence and honesty for his income and return on capital, while the farmer had less incentive to innovate and exert himself as he only received half the crop. The result, as the Physiocrats lamented, was a vicious circle of poverty, underinvestment and low yields, which undermined French society and the nation. A.R.J. Turgot, as *Intendant* of the Limousin, drew a picture of the *métayer* complacently growing black corn and chestnuts, instead of wheat and more advanced crops, safe in the knowledge that his landlord would support him for fear of the land passing out of cultivation. The landlord, equally trapped, 'often far from rich, at the mercy of a thriftless tenant, bearing all the risk and being entitled to only half the produce, did not think it worthwhile to invest his small capital'.[3] In the literature of the French Revolution, *métayage* epitomized the *ancien régime*. Its reputation as a bad tenure was sealed by the Marxist historians, Bloch and Goubert. According to Bloch, the sudden increase of *métayage* in the sixteenth century had reversed the trend towards the emancipation of the peasantry by allowing landlords to reassert feudal control through a capitalist market, impoverishing whole regions of France as a result.[4] In Goubert's rural hierarchy the lowly *métayer* occupied the unenviable position between farmer and labourer ruthlessly exploited and ensnared by debt, ignorance and poverty.[5]

Not surprisingly, *métayage* and other forms of sharefarming have attracted little interest from historians of English history. This was a French problem for French historians, a view reinforced by Hilton's article 'Why was there so little champart rent in Medieval England?', which is the only full discussion of the issue in an English context.[6] No mention of sharefarming appears in the indices of *The Agrarian History of England and Wales*, nor in the text on English agricultural history by a co-author of this book.[7] In his recent contribution to the history of landlord and tenant relationships in eighteenth-century England, Stead cites with approval Adam Smith's dismissal of sharecroppers: 'They have been so long in disuse in England that at present I know no

English name for them'.[8] The general belief is that sharefarming did not exist in England or at least was of such rarity to be of minimal significance.

However, evidence has come to light that sharefarming did exist in England from the middle ages to the present day, at different levels of society and throughout the country. But unlike the situation in France, English sharefarming was never institutionalized, remained largely informal and rarely appears in documents. When it does appear, historians, unfamiliar with the practice, fail to recognize its significance. In this way, a stratum of farming activity, embracing collaboration between farmers and landowners, has been lost. The aim of this book is to reveal that hidden history of sharefarming.

The possibility of accomplishing this task originated in the discovery of letting to halves agreements in the estate book of William Windham of Felbrigg in Norfolk.[9] These sharefarming contracts bear a close resemblance to the examples cited by Goubert.[10] Between 1673 and 1688, Windham entered into several such agreements, which invariably involved leasing a herd of dairy cows to the tenant and sowing the arable to halves: the tenant paid a fixed rent for his cows, but shared the cost of cultivation with the landowner. At the end of the year, they divided the crop of grain. In this way, Windham provided working capital to the tenant and shared the risk of farming. During this period of sharply falling corn prices, it enabled him to diversify into stock farming and ease tenants into vacant farms. Subsequently, further well-documented examples were found on other leading Norfolk estates, including Raynham and Hunstanton. At Raynham, home of the Townshend family, 40 per cent of the estate was either farmed in hand or let to halves in the 1690s by the trustees for Charles, 2nd Viscount Townshend, famously known as 'Turnip Townshend'. His father, Horatio, brother in law of William Windham, had also used the device in various ways to attract tenants since the 1660s. At Hunstanton, where the Le Stranges drained hundreds of acres of coastal marshes in the 1630s and 1640s, they resorted to farming to halves with tenants on newly cultivated marshland. Thus, farming to halves was used by Norfolk landowners not only to counter adversity in the depressed conditions of the late seventeenth century, but to expand and intensify production in the early part of the century when prices remained high. Throughout this century of economic and political turbulence, landowners used the practice in a flexible and informal way to support undercapitalized tenants and mitigate the risks of farming.

These Norfolk agreements provide a firm historical base, facilitating comparisons with other parts of the country, with other times and with different situations. Farming to halves was not a purely Norfolk phenomenon, neither was it confined to the seventeenth century, nor was it peculiar to large landed estates. Historians, now alert to the practice, have identified agreements in their records, which hitherto they had overlooked, while references in well known published accounts have come to the surface, confirming that forms of sharefarming were widespread in peasant communities throughout Britain.[11] Robert Loder's farm accounts show farming to halves supporting widows in their old age, allowing yeomen to expand their enterprises, and enabling labourers to secure a foothold in farming.[12] Farming to halves thus existed at two distinct social levels. Between landlords and tenants when times were hard, situations difficult or when tenants were unwilling to pay fixed rents; and amongst the peasantry where it was used to overcome a temporary difficulty and to ease the flow of capital and credit. In this way, sharefarming in England occurred vertically and horizontally in rural society, and in doing so conforms to the European pattern.

These findings are significant. First, they affect our understanding of the development of English agriculture. Economic historians have long identified the English route to agrarian capitalism as one whereby complicated forms of medieval tenure and property rights were replaced with a system of private ownership, in which most of the land of England was farmed by capitalist tenant farmers paying rack rents to their landlords, and worked by wage labour.[13] Whereas feudal institutions and communal cultivation had involved spreading the risk of farming throughout the village community, the modern landowner exercised complete control over his capital and received a guaranteed income, while the tenant bore the risks of farming and profited from his efforts. The existence of sharefarming disrupts this supposedly simple dichotomy. If, in some circumstances, landlords were providing working, as well as fixed capital then some modification is required to the most recent views on the relationship between landlord and tenant and the balance of risk in farming.[14]

Sharefarming also complicates the meaning of rent in an historical context. We know that payments from tenants to landlords before the nineteenth century should not necessarily be interpreted as rent in any modern sense.[15] For various reasons they often bore little relation to the commercial value of the holding, but were the legacy of feudal contracts and an almost endless variety of customary agreements. For example, payments from tenant to landlord could represent the value of former labour services, which may have been fixed by custom at medieval values. These difficulties are compounded if sharefarming agreements were in place, but hidden behind more conventional looking rents. Some payments may be for the leasing of animals, rather than land, while other payments could be considerably less than the actual rent, if the landlord received half the crop.[16]

The hidden nature of sharefarming in England is also of particular significance to the debate on the development of economic institutions. Ogilvie argues that it was not necessarily the case, as economists have claimed, that formal institutions were the most efficient solutions to economic problems.[17] These institutions were often the result of negotiation between the principal parties, which could inflict serious damage on third parties. The English landlord-tenant system, from the perspective of landowners and tenants was an efficient institution, but small farmers and wage labourers, who were excluded from the discussion, paid a heavy price. Exclusion led to the creation of black markets and shadow economies with their own informal institutions, which often underpinned or supplemented the working of more rigid formal institutions. Medieval historians now recognize that subletting, hidden behind manorial records, performed just such a role. We would argue that sharefarming in England also fits this category, and needs to be incorporated into our analysis of tenure as an institution.

The existence of sharefarming in England also has far wider implications since it connects us more closely to developments in Europe. In the past, historians have emphasized the differences between the experience of English peasants and their European counterparts, but now they stress the similarities; the discovery of *métayage* in Britain has the potential to transform that debate. At the same time, the status and role of *métayage* in European agriculture has been radically revised. No longer is it regarded in Goubertian terms as a bad tenure, but a flexible and efficient response to difficult markets and oppressive tax regimes. The consensus now is that *métayage* was the consequence of poverty, not the cause of it: taxation debilitated French farmers, not *métayage* tenure. Recent research associates *métayage* with enlightened stock farming and areas which witnessed the most rapid rise in productivity in France under the *ancien régime*; a picture not dissimilar to the experience in Norfolk.[18] These findings question the link between the efficiency of farming and the tenurial regime, which was of such concern to the eighteenth-century agriculturists and political economists.[19]

Revisionism is not confined to France. Galassi, who writes on sharefarming in Tuscany, believes that its abolition in the 1950s, on wholly political grounds, has proved a tragedy for Italian agriculture. *Mezzadria* had been the mainstay of Tuscan agriculture for a millennium – Galassi quotes contracts dating from the ninth century. These agreements were particularly favoured by urban landlords in the thirteenth century as the most efficient way of ensuring food supplies for growing populations. Significantly, in the aftermath of the Black Death, sharefarming was developed further as it was the ideal arrangement to effect diversification into high quality food production, notably wines, olive oil, fruit and wheat. It was infinitely flexible, provided sufficient control for the landlords and enough incentive to the operator. Galassi's model shows that in 1427, 97 per cent of Florentine vineyards and fruit orchards owned by urban landlords were cultivated in this way and that these types of contract continued until the twentieth century, 'a proven success over the long run'.[20]

The outlawing of sharefarming in France and Italy after the second world war is now seen as a political response to the authoritarian regimes of the 1920s and 1930s, notably the neo-feudalism of Mussolini's Italy.[21] This view is not unconnected with the widespread disillusionment with the Common Agricultural Policy. Today the state is regarded as just as exploitative and far less efficient than feudal lords or indeed Fascist dictators, when it has tried to protect farmers' interests and compensate for defective markets.[22] In developing countries, sharefarming is no longer discouraged on narrow ideological grounds, but accepted as a solution to many of the problems encountered at primitive levels of production: the lack of credit, availability of skilled labour and the inherent risks of farming. In technical terms, as Galassi argues, sharefarming reduces transactions costs, avoids the moral hazard of hiring and supervising direct labour, and motivates the workforce by providing incentives, while the crop provides the collateral for loans. The wheel has turned full circle as this shift of opinion is reflected in the latest policy initiatives from the European Union which increasingly promote profit-sharing schemes and partnerships in farming.

The most powerful case for modern sharefarming has been made by the outstanding success of the unsubsidized agriculture of New Zealand, where share contracts predominate. It was through this route that sharefarming first attracted the attention of English landowners in the 1980s. At that time, sharefarming offered a way round the Agricultural Holdings Act (1976), which extended security of tenure to three generations. In these circumstances, share contracts avoided the creation of a tenancy.[23] These restrictions on landowners were removed by the Farm Business Tenancy Act (1995) and for a time interest waned. The figures produced by the Ministry of Agriculture, Fisheries and Food (MAFF) in 1997 showed a slight decline in the numbers of holdings sharefarmed in the previous two years. Since then, with the collapse of agricultural incomes and the worrying plight of many rural communities, collaboration is increasingly viewed as the way forward by landowners, tenants, agriculturists and politicians. The manner in which New Zealand successfully weaned itself off subsidies in the late 1980s, with benefits to the entire economy, has provided important lessons. In that process, the prevalence of family farms and the flexibility of sharefarming was a critical factor enabling farmers to respond to fluctuations in income and lower returns on capital. It is no coincidence that the dairy industry in New Zealand, with its tradition of sharemilking for new entrants, emerged as the most buoyant sector. In this context, sharefarming in Britain has become a mainstream alternative to more traditional tenures, yet there remains a general view that it is not, and never was, a British way of doing things. This book will dispel that myth.

The book begins with a short chapter on the theory of sharefarming and the place it occupies in agricultural economics. How do economists explain the continuing use of sharefarming, and indeed its revival in the twentieth century, despite its condemnation by the great political economists of the eighteenth and nineteenth centuries? What do we mean by sharefarming? Sharefarming in England, rooted in local custom, existed in many different forms and was known by different names in different parts of the country. This has added to the general confusion and lack of awareness. To clarify our understanding of the practice, a taxonomy has been constructed to explain the various types, from farming to halves, the form that most resembled *métayage*, to profit-sharing, incentive payments, animal leasing and payments in kind.

The chapters that follow are arranged chronologically, but where to begin is the first difficulty. The origins of the practice cannot be precisely identified, so we start in Chapter 3 with the first indicators as they arise in medieval documents. This is not easy as English medieval records rarely contain explicit references to farming to halves as they did not constitute the transfer of an estate in property and were thus of little interest to landowners. It sometimes appears when restrictions against subletting were punished in the manor courts, but finding examples is purely a matter of chance. So, we are forced to rely principally on secondary sources, scouring accounts for evidence, as farming to halves is seldom indexed. References are more common than one might suppose, but a particular question is why leading historians, aware of the shortcomings of manorial records, the prevalence of subletting and the existence of sharecropping, have not speculated more fully on the role this type of informal agreement might have played in village communities and over the lifecycle of a peasant family. The chapter concludes with a section on the variations that existed throughout the British Isles, notably in Scotland and Wales where it was more fully documented, which provides a context for the English practice.

In the early modern period, evidence of farming to halves comes to light in a range of documentary sources, including probate inventories, farm accounts, estate records and legal documents. More examples were found for this period, particularly the seventeenth century, than for all other periods. Chapter 4 considers whether this is simply the result of better documentation, or whether there were particular reasons why farming to halves was more common during this century, so strongly affected by fluctuating prices, economic turbulence and political instability. Did the prevalence of sharefarming at this time represent a crucial phase in the transition to agrarian capitalism, just as in other parts of Europe? If so, why does it not figure more prominently in the accounts of early modern historians?

Chapter 5 presents case studies of sharefarming in the seventeenth century from four Norfolk estates. These provide conclusive archival evidence that sharefarming was used systematically in England by leading landowners less than a century before Arthur Young denied its existence. The chapter is divided into two sections to emphasize how landowners adapted the practice to suit different situations; first as a strategy for expanding and intensifying cultivation in the optimistic years before the Civil Wars, and secondly as a way of responding to falling prices from the 1660s. The estate records indicate that the practice ended as landowners simplified their approach to estate management and let large farms to well capitalized tenants at reduced fixed money rents. But, glimpses of sharefarming on these estates and elsewhere point to its survival between tenants and amongst more lowly farming circles. By this time landowners had found more acceptable ways of farming their land, while amongst the peasantry sharefarming continued.

Chapters 6 and 7 explore more fully what happened to farming to halves in the eighteenth and nineteenth centuries, from 1720 to 1900, when no evidence of sharefarming has been found in the documents. Chapter 6 begins by showing the survival of farming to halves on the Norfolk estates in the early eighteenth century, contrary to the impression gained from the records of those estates. A particular feature was the subletting of dairies and the leasing of dairy cows, which survived in parts of England until the twentieth century. Evidence of cow leasing will be examined, using farm diaries and the eighteenth-century agricultural surveys. Finally, these findings will be compared with the most recent research on farm tenures in England and sharecropping in Europe.

Chapter 7 examines the literature on land reform and shows how a form of sharefarming re-emerged in the late nineteenth century as a means of ameliorating the plight of agricultural labourers. This was profit-sharing, rather than the produce sharing of the seventeenth century, a device designed to provide incentives to wage labourers, not to share the output between landowner and farmer. The new version illustrates the flexibility of the practice and how it could be modified to suit changing circumstances and different situations. This chapter uses the later agricultural surveys, the reports of the agricultural commissions and the writings of philanthropic landowners who engaged in profit-sharing, to show that forms of sharefarming attracted much attention at this time. Comparisons will be made with colonial New Zealand where sharefarming was adopted enthusiastically during this period and where it proved so successful.

In Chapter 8, the return of farming to halves in the early twentieth century confirms what we suspected in Chapter 5, that the practice survived in its original form amongst tenants, yeomen, smallholders and landless labourers as a traditional way of sharing the risk of farming and supporting each other in times of difficulty. Farm diaries, newspaper advertisements and oral evidence from farmers and landowners, show how the practice worked, particularly in the Welsh border counties and in East Anglia during the course of the twentieth century. The impression gained is that the practice never died out, but simply remained secret, concealed and hidden from view. From the landowners' perspective, we will see how sharefarming came to the notice of the agricultural establishment in the 1980s, through the introduction of the idea from New Zealand and its promotion by the Country Landowners Association. The scale and success of the practice will be assessed through interviews with landowners, agents and contractors. Finally, Chapter 9 brings these issues up to date, by using a contemporary survey to measure the extent of sharefarming in England at the start of the twenty-first century.

Essentially, the aim of the book is to reveal a lost history and to show how that lost past relates to the present. In these hard-pressed times, as Thirsk writes, historians have a particular role: people want to know how their forbears coped in periods of adversity. It is our task to shed light on that process.[24]

2
Sharefarming in England: Theory and Practice

Whilst not a paramount concern amongst historians, sharefarming has excited much controversy and generated a far-reaching debate amongst economists; they are interested primarily in how it works, why it exists, and why, despite the disapproval of the classical economists and the emergence of more satisfactory systems, it has survived.[1] The purpose of this chapter is to not to survey the entire academic literature, but to explain the broad principles of sharefarming and to construct a taxonomy from which to view the empirical evidence that follows. Sharefarming exists in infinite variety. The taxonomy leads with the classic version of sharefarming, which involves the sharing of inputs and outputs, and moves on to consider other forms, which include the provision of working capital to tenants, the paying of corn rents and profit-sharing with labour. These do not strictly qualify as sharefarming, and more correctly should be termed unconventional tenures to distinguish them from conventional fixed rent tenancies. However, with all these agreements landowners shared the risks of farming with the tenant, whether sharing the crop, sharing the profit on livestock, risking his herd of cows or taking a gamble with the price of corn in the market. It is this direct involvement which differentiates forms of sharefarming from fixed rent tenancies, and elicited the disapproval of the political economists.

The debate on sharefarming, which dates back to the eighteenth century, turns on the assumption that sharecropping, with its shared inputs and outputs, is an inefficient and irrational form of tenure. It was condemned wholeheartedly by the political economists, who regarded *métayage* as a major obstacle to improvement. As Adam Smith explained:

> It could never be the interest of this last species of cultivator [the *métayer*] to lay out in the further improvement of the land, any part

of the little stock which they might save from their own share of the produce, because the lord, who laid out nothing, was to get one-half of whatever it produced. The tithe, which is but a tenth of the produce, is found to be a very great hindrance to improvement. A tax, therefore, which amounted to one-half must have been an effectual bar to it. It might be the interest of the métayer to make the land produce as much as could be brought out of it by means of the stock furnished by the proprietor: but it could never be his interest to mix any part of his own with it. In France, where five parts out of six of the whole kingdom are said to be still occupied by this species of cultivators, the proprietors complain that their métayers take every opportunity of employing the masters' cattle rather in carriage than in cultivation; because in the one case they get the whole profits to themselves, in the other they share them with their landlord.[2]

Marshall endorsed this view, underlining the fundamental inefficiency of a system that reduced output below the maximum feasible. 'If he [a French *métayer*] is free to cultivate as he chooses, he will cultivate far less intensively than on the English plan'.[3] In modern terms the sharecropper has an incentive to supply less than optimal labour because he receives only a part of the increase in production from an additional unit of labour. As Smith said, the landlord's share acts as a tax, which reduces the tenant's incentives. Thus sharecropped farms are bound to produce less than the maximum possible, which Marshall assumed was achieved on capitalist farms leased on fixed rents, as in England.

If the system was so inefficient, why did landlords and tenants choose it? One of the principles of institutional economics is that the rational and informed participants will adopt the output-maximizing solution.[4] So what made them select sharecropping? This is known as the 'Marshallian Puzzle'. Over the years, economists have offered various explanations: sharecropping reduced the monitoring costs of wage labour by offering incentives to the tenant; it allowed the parties to share the risk of fluctuating farming incomes, and supplied poor tenants with working capital.[5] These factors often outweighed the advantages of wage labour, which was more flexible but needed supervision, and of fixed rent tenancy, where the entire risks of farming were assumed by the tenant in return for the entire output. While most economists accept the undersupply of labour with sharecropping and the tendency to under-report the output, they also point out the risks attached to fixed rent tenants. As they received the whole crop, these tenants had the incentive to over-exploit the land, which could cause permanent damage to the farming environment. Neither contract was optimal, but a rational landlord might in certain situations prefer to sharecrop in order to minimize the risk of long-term losses. This was particularly true of Mediterranean countries where cultivation often involved expensive and delicate crops such as vines and olives over which the landlord wished to keep control.[6] Tenants on fixed rents could also miscalculate, amass huge debts and throw up their tenancies, with permanent loss to both landlord and tenant capital. In these circumstances, as farms lay vacant, it was often more practical and profitable to let farms to halves than to cultivate them with direct labour.

There is no shortage of examples to show that in given situations sharecropping was, and remains, the most rational decision for landowners and farmers.[7] This has become increasingly evident in Third World economies where the crop share motivates the workforce, provides collateral for loans, and helps sharecroppers accumulate capital.[8] A more difficult problem is to define precisely what is meant by sharecropping, which unlike fixed rent tenancy, is not a clear-cut relationship between landlord and tenant. As Federico points out, the division of the product between the parties according to some pre-determined share-out leaves the door open to a wide range of sub-types.[9] The key feature is that the income of both landlord and tenant depends on the output of the farm. This differs from a fixed rent tenancy, where only the tenant's income and profit is related to output. Similarly, where a sharecropping or profit-sharing relationship exists between a tenant and labourer, it differs from standard incentive contracts like piece work, which relate the worker's remuneration to his own performance, by linking the labourer's income to the overall output of the farm. The tenant's share of the product can be as high as 80 per cent or as low as 20 per cent, depending on a range of factors: the provision of capital, the division of expenditure, the degree of tenant's freedom, the nature of the crop and so on. In practice if the proportion is too low, the sharecropper can be worse off than a waged labourer, and the relationship highly exploitative.[10] The terms of these contracts are endlessly flexible and can be modified indefinitely to suit changing circumstances and new requirements. The most recent examples from Madeira, Spain, Portugal, France and Italy emphasize the positive nature of sharecropping, how it was used to effect improvement and diversification into new crops, and how it often co-existed with fixed rent tenancies.[11] In all these cases sharefarming evolved over time and space in response to a wide range of economic, social and institutional factors.

The recent emphasis on evolution, variation and flexibility helps us to understand the development of sharefarming in England. However, as sharefarming was never institutionalized in this country and we have no general name for it, there is a particular need for a taxonomy to explain the terms. The classic version is known as farming, sowing or letting to halves, but also as halfendeal, (half the deal) in Cornwall, the partible way in Norfolk, *champart* in medieval documents, and half crease (half the increase) on livestock in the west country and Welsh border counties. A further array of terms can be found in other parts of Britain, the most common being steelbow tenure in Scotland. In addition, there were stock and land leases, cowleases, corn rents and corn agreements which differ from the classic versions in that they do not involve a share of the output. Sometimes, the difference is very narrow, but all schemes, to a greater or lesser degree, were designed to mitigate the risks of farming. Tenurial agreements are essentially a negotiation between the owner of the land and the producer, but only an extremely confident tenant will agree to pay a fixed rent for a number of years, and provide all the working capital. As Adam Smith pointed out, the conditions need to be optimal, predictable and generous. These conditions did not always apply and other sorts of agreements were needed in different places, at different times, for different reasons, and for different purposes. The taxonomy in Table 2.1 is based on our findings and remains provisional and descriptive in character. Each type has been given a letter, or a key, which cross-references to the text and the county survey in Appendix 1.

A significant omission from the taxonomy is the tithe, which is sometimes cited as an example of English sharefarming.[12] Indeed, as we have seen, Adam Smith compared *métayage* with the payment of the tithe in England, which he also regarded as a hindrance to improvement. In Scotland, 'thirds' of the crop were often collected with the 'teints' [tithes], so the association is understandable, if erroneous.[13] The tithe was essentially a tax imposed by the clergy on their flock, and not a tenurial or farming agreement. A clergyman might well enter into a sharefarming agreement with a farmer over the cultivation of part of the glebe, or the grazing of livestock, but this was entirely different from the payment of the tithe by the whole parish. As it happens, the Le Stranges of Hunstanton, who were substantial lay impropriators, collected their tithe corn with their farm rents paid in corn, and corn sown to halves, illustrating the flexibility and variety of sharefarming and unconventional agreements.

Table 2.1   A taxonomy of sharefarming in England

Sharefarming type	*Key*	*Terms and descriptions*
Sharing the output	A	Farming to halves, sowing to halves, letting to halves, involving a division of the crop instead of a fixed money rent. Classic form of *métayage* tenure. Also known as farming 'the partible way' in Norfolk, halfendeale, or 'half the deal' in Cornwall, and in medieval documents as *champart* rent. Paying a portion of the lambs to the shepherd. Placing ewes at halves: half the wool and lambs. Steelbow tenure in Scotland, often collected with tithes, 'teints and thirds'; a third of the grain crop. Also known as half-foot, or leth-cas, with livestock.
Sharing the profits	B	'Half crease' – half the increase or uplift on fatstock, common in the Welsh border counties and Cornwall.
	C	Profit-sharing with labourers, regulated by 'enlightened' landowners in the 19th century.
	D	Modern share and contract farming, joint ventures between landowners and contractors.
Partnerships, sharing resources and crops	E	Farming partnerships, halving flocks, half shares in stock, grain and equipment, often found in probate inventories as moieties or halves, notably in Cornwall.
Providing working capital	F	Stock and land leases: ancient arrangement of leasing farms with livestock
	G	Cow leasing: dairy herds leased to dairymen, common in Dorset and the west country.
Corn Rents	H	Dating from medieval times, consisting of fixed amounts of grain often paid with money rents.
Corn Agreements	J	Prices guaranteed by the landowner, in effect granting a subsidy to the tenant.

A. Sharing the output: farming to halves

The first group includes those agreements which involve sharing the input and output of the farm. This undisputed version of sharefarming in England is known as farming to halves. As we have seen, in the documents it appears in slightly different forms: 'sowing to halves' with grain, 'letting to halves' by the landowner, and when it was applied to livestock, putting or placing animals, mostly ewes, 'at halves'. In these classic arrangements, the landowner provides the land, stock, seed and equipment as appropriate, the tenant supplies his labour, and they share the output, either in grain, or lambs and wool. No money changes hands and the deal is settled at the end of the year, removing the need for external borrowing and credit. In this internalized system, tenants build up working capital, and the landowner receives some security over his investment in the form of the crop or livestock. Why did landlords and farmers enter into these agreements? The common denominator seems to be the association with risk. The most detailed examples of sowing to halves were found on the very light soils of west Norfolk, where grain growing required the most careful cultivation and substantial investment in sheep flocks to manure the land. However, the well-developed grain market in East Anglia made the risk worthwhile and encouraged landowners and tenants to collaborate. The difficulty was in finding tenants with sufficient capital to lease and stock the flocks, and take the risk of cultivating the poor arable land. Thus landowners, in order to maintain this delicate farming regime, sometimes divided flocks and let the arable to halves.[14] Similarly, the practice of placing ewes at halves was found principally in the Welsh border counties, noted for their lawlessness and consequent risks to livestock. In the sixteenth century, the rising demand for fatstock and stores, and the union with England, persuaded English dealers to share the risks with the isolated Welsh hill farmers.[15] In Scotland, where farming to halves was known as steelbow tenure, all manner of permutations existed reflecting the less favourable environment of that country, and the military nature of this tenure designed to protect border farmers.[16] In the Highlands, agreements with crofters remained paternalistic in motive, while in the improving Lowlands, many agreements by the eighteenth century were highly exploitative of labour, reflecting landowners' desire to capitalize on the growing grain market with England.

The proportion received by the landowner, or kept by the tenant, varied depending on the input of capital, the geography and the risk borne by the tenant. For example, at Wolverley, Worcestershire in 1230, tenants paid half the crop for demesne land that was manured, and a third for land that was not.[17] In Gaelic Ireland, where tenants supplied the seed and equipment, they paid a quarter of the crop as rent, retaining the rest or 75 per cent.[18] These arrangements can be compared to the graduated scheme that operated in New Zealand until quite recently. A sharemilker might start farming in return for 29 per cent of the milk cheque supplying just his labour, progress to 39 per cent when he provided the tractor and trailer, and finally to 50 per cent when he supplied the cows. With the steeply rising value of land in New Zealand a sharemilker today is lucky to secure a 25 per cent agreement.[19] This demonstrates the variable nature of sharefarming agreements and how they are continually modified to suit changing circumstances.

B. Sharing the profits: half crease

This type is closely aligned to farming at halves, but involves the division of the profits, rather than the output, in other words, money changes hands. Most common is the term 'half crease' or half the increase, which involves halving the uplift or profit on fat cattle, let out to the tenant on a grazing agreement. With suckler cows it might involve halving the calves, but this was much less frequent than halving the lambs and wool from ewes. Half crease existed alongside the practice of placing ewes at halves and livestock 'at tack'. With tack, farmers paid a fixed price for each animal for a defined period and did not share the product or the profits with the grazier.[20] References to half crease were found in the Welsh border counties and in Cornwall, where it continued until the 1970s.[21]

C. Profit-sharing with labourers

The objectives of profit-sharing with labourers were very different from sharing the profits. It was based on a French idea, and was introduced by enlightened landowners in the late nineteenth century as a way of encouraging labourers to stay on the land.[22] Typically, labourers were paid a wage, landowners received a rent and the surplus was shared according to a formula devised by the landowner. In addition, the landowner might provide a cow with pasture and feed for which the labourer would pay a modest sum per week. Despite the exhortation of philanthropic landowners, the scheme did not fit easily with the landlord-tenant system, where three parties were involved. It worked best when estates were farmed in hand, and managed by a bailiff who also received a share of the profits. Another form of incentive was to pay skilled labourers, notably shepherds, a share of the lambs and wool. This happened at Raynham in the 1620s and 1630s when the Townshends were building up the size of their flocks.[23] It also occurred in Australia in the 1840s, when convict labour dried up and skilled shepherds demanded a portion of the lambs and wool.[24] With sheep farming, a proportion of the lambs and wool was a simple share to calculate, which might explain its long-term use and popularity.

D. Modern forms of contract and sharefarming

Modern forms of sharefarming in England most closely resemble the profit-sharing schemes of the late nineteenth century in the way that they provide for guaranteed payments to landowner and contractor, before the surplus is divided.[25] The agreements, provided by the land agency firms, set out the role of farmer and contractor, specifying the provision of capital made by the farmer: land, building, fixed equipment and breeding livestock; and by the contractor: labour, machinery, management expertise, and livestock if not provided by the farmer. The financial arrangement is that all costs applicable to the agreement are deducted from the total output of the enterprise. Fixed costs include the 'Contractor's Basic Fee' and the 'Farmer's Prior Charge'. What remains is 'The Divisible Surplus', with the ratio between the farmer and the contractor often fixed in the contractor's favour. More informal arrangements still exist and operate on the basis of a handshake; these partnerships rely heavily on trust and are designed specifically to avoid the attention of the lawyer, the taxman and the agent.

E. Partnerships and sharing crops and resources between farmers

The most difficult type to categorize are the half shares and moieties which appear in probate inventories from the sixteenth to the early eighteenth centuries. Sometimes these entries are accompanied by references to halfendeal, half crease, or partible wheat, barley and oats, but it is not always entirely clear whether the half shares reflect a sharefarming agreement. Half shares sometimes extended to household goods and farm equipment indicating a full partnership, rather than a short-term informal farming agreement. Certainly, Norfolk landowners encouraged partnerships in the seventeenth century, dividing flocks and leasing them to syndicates of farmers.[26] This may have happened when large single tenants were not forthcoming, but equally landowners may have preferred the security of numbers and sharing the risk amongst several tenants. For example, in 1605, Sir Hamon Le Strange leased the foldcourse at Barrett Ringstead to six tenants and sowed the arable to halves. The practice continued in the late seventeenth century as grain prices fell and tenants needed support.[27] Partnerships and the sharing of resources remind us of medieval open field farming, where tenants farmed scattered strips and worked within the framework laid down by the manor court in order to mitigate risk and to provide for the subsistence of village communities. The evidence from probate inventories indicates a continuation and modification of these arrangements in a period of increasingly commercial agriculture.

F. Providing working capital: the stock and land lease

The stock and land lease, like the tithe, has long been associated with *métayage* tenure, in that it supplies working capital, in the form of stock, to the tenant.[28] But critically these leases did not involve sharing the output of the farm or the uplift on the stock; the tenant had to take his chance in the market and reinstate the stock at the end of the term. It was an ancient tenure, whereby the landlord leased stock – cows, sheep, horses, oxen, pigs – with the land to enable the tenant to start farming. The health of the stock always presented a difficulty, requiring the landlord and tenant to share the responsibility for different diseases. Leases petered out in the mid-fifteenth century, as landowners permanently withdrew from demesne farming. But the principle of landowners helping tenants with stock persisted, for example, Charles 'Turnip' Townshend, made loans to tenants to purchase sheep when the flocks, previously in hand, were leased in the early 1700s.[29]

G. Cow leasing

Cow leasing was a system whereby the landowner, or farmer, would lease a herd of cows to the tenant or sub-tenant.[30] The practice dates back to the middle ages, where the 'lactage' of each cow might be leased for 5s a year, compared with 45s in the 1670s.[31] In the late seventeenth century, it was used alongside sowing to halves to support failing tenants and effect diversification into dairy farming, and continued as a mainstay of the west country dairy industry until the nineteenth century.[32] It was the precursor to modern sharemilking, a system made possible by the development of farming co-operatives in the 1890s, which provided for the division of the monthly milk cheque between landowner and dairyman removing one of the principal problems of monitoring milk production. Cow leasing, with landowners assuming responsibility for herd replacements, involved considerable risk and certainly qualifies as a risk sharing venture, even if it was not strictly sharefarming.

H. Corn rents

Corn rents do not constitute sharefarming, being a fixed payment, albeit in kind. But in his receipt of grain, instead of money, the landowner was in fact sharing the risk of the market. These rents were common in Norfolk in the fifteenth century and survived into the eighteenth century.[33] At Sedgeford, the Le Stranges collected the corn rent of two bushels per acre, with the tithe corn and some money rent. Landowners contributed storage facilities and management skills, shared low prices, and benefited from high prices in times of dearth. Even in the middle ages, these commercially based corn rents were popular, as landowners alternated between direct farming and leasing the demesne.[34]

J. Corn agreements

Corn agreements differ from corn rents in that they represent a direct subsidy on corn production with the landowner buying in corn at a guaranteed price, similar to those provided by the European Union. Landowners shared the fluctuations of the market, but avoided sharing the risk of farming with the tenant. Sir John Hobart entered into corn agreements in the 1670s and 1680s, when the Townshends and Windhams were sowing their arable to halves.[35] No evidence of corn agreements was found elsewhere, but they should be seen in the context of the Corn Bounties paid by the government to farmers, during this period, to encourage the export of grain, the idea being to support rents and taxes by keeping arable in cultivation.[36] A modern variant involves the contractors buying the crops at a predetermined price, and taking a gamble on the market.

The taxonomy illustrates the variety of forms of sharefarming and unconventional tenurial practices that have existed in Britain since at least the middle ages. They differ fundamentally from the fixed rent tenancies, or landlord-tenant system which has been considered the norm. Sharefarming existed at quite distinct social levels: between landlord and tenant, tenant and sub-tenant, amongst the peasantry, and even between farmers and labourers. Sometimes, it existed on a single holding where the large tenant paid a fixed rent and sublet portions to halves. In all these contexts, it performed slightly different functions, but was principally concerned with mitigating risk, easing the flow of capital and credit to farmers, securing tenants and providing incentives to labour. Given the informal and often covert nature of these agreements the evidence is difficult to quantify, and the extent of sharefarming in England remains an enigma. However, farming to halves was clearly more common in certain areas and at particular times. In the following chapters an attempt will be made to explain more fully why this happened, and in the process, a chronology of the occurrence of sharefarming in England will emerge.

3
Sharefarming before 1500:
A Hidden Practice

Where to begin a history of sharefarming in England? If the definition of sharefarming is confined to the classic versions, involving the sharing of the input and output, the task is beset with difficulty. Medieval documents, which provide such a precise record of the transfer and ownership of land, are more or less silent on farming to halves, as it did not involve a transfer of ownership or obligations.[1] Manorial court rolls, surveys, estate rentals and accounts that were designed to inform the lord of the rents and services owing to him from his tenants, say little about their farming and do not necessarily reflect the actual occupancy of a holding or how it was used. However, formal leases often contained restrictions against subletting, and it is in this context, when infringements were punished in the manor courts, that evidence of farming to halves comes to light.[2] These formal leases included stock and land leases which were of great antiquity.[3] Lennard dates them back to the Anglo-Saxon period, whilst Venn describes *métayage* as the most ancient tenure of all, and in its original conception exactly akin to the English stock and land tenure.[4] Broderick, in his nineteenth-century critique of the landlord-tenant system, compares the stock and land lease to the *métairies* of south west Europe and recommends its revival in England.[5] Thus, if we extend our definition of sharefarming to include the stock and land lease, generally considered by historians and economists to be the English version of *métayage*, we can see that forms of sharefarming date from the earliest times and most likely predated fixed money rent.[6] The difference between the stock and land lease and farming to halves lies in its legal status, which ensured that the former was recorded, while the latter seldom appeared in the documents.

The difficulty of documentation is compounded by the nature of farming to halves. Like subletting, it was essentially an arrangement that peasants would wish to conceal as far as possible from manorial officials for fear of alerting them to the true value of their holdings and to new ways of raising manorial income. Even today, it is used as a device to circumvent the law and to avoid the taxman. In other words, the evidence is deliberately concealed and intentionally elusive. The task is made even more difficult as farming to halves was known by different terms in different parts of the country and took slightly different forms, a reflection of its covert nature, lack of legal status and origins in local practice and custom.

Perhaps the most intractable problem is that historians have not associated sharefarming with the English experience. Many are unfamiliar with the practice and fail to recognize it in the documents. Even when they do mention it in their work it is seldom analysed or indexed. Some appreciate that the practice existed and accept that it was probably the most common agreement between peasants, yet few have drawn sharecropping, as it is usually termed in medieval accounts, into their description of English peasant life, despite its prevalence in Europe.[7] Campbell comes close when he describes the methods employed by tenants as they sublet their holdings on short-term leases, and refers to 'small amounts of land let for a single year or crop, often on a "share-cropping" basis, whereby the lessor supplied the seed and working capital and in return received labour and a share of the crop', yet he does not question or explain the practice, or consider how it might help to answer some fundamental questions posed by medieval historians.[8]

Thus both the documents and the secondary material conspire against us. Searching the archives for such elusive evidence was clearly out of the question. However, armed with an understanding of sharefarming, progress was made by trawling through the secondary material, following leads and alerting medieval historians to the practice. This chapter begins by focusing on farming to halves *per se*, (Types A and B in our typology) and analysing the references found in the work of leading medieval historians.[9] No apology is made for our reliance on secondary sources, in fact, showing how the practice has been approached by historians helps to explain why sharefarming has remained hidden from view. Drawing on the extensive literature on the lifecycle of a peasant family, we then consider how farming to halves may have worked in village communities. The chapter concludes with a survey of the variations that existed across the British Isles, developing more fully the points made in the taxonomy. This is by no means a definitive account. The primary aim is to draw attention to a hidden practice, so that historians can identify it in the documents and appreciate the role it played in the development of medieval rural communities throughout the British Isles.

A history of English sharefarming must begin with Hilton's article, 'Why was there so little champart rent in Medieval England?', which is the only full discussion of sharefarming in an English context and has largely shaped perceptions of the practice in this country.[10] As the title suggests, he considers '*champart*', another name for sharefarming derived from the Latin *ad campi partem*, to have been relatively unimportant in England, but significantly he does not deny its existence. His purpose is to explain why it formed a limited element of the rents paid to English landowners, and why they came to prefer money rents and labour services, in contrast to their European counterparts. His article was first given as a paper to a French conference and should be regarded in that light; it was not designed for an English audience and was only published in the *Journal of Peasant Studies* several years later.

Hilton offers a definition of *champart* rent and a chronology of its occurrence in the medieval period. He is not concerned with food rents or fixed rents in kind – the corn rents in our taxonomy (Type H) – but specifically with those rents in kind calculated as a fraction of the total product of tenanted land, which related to the French system of *métayage*. He dates the existence of *champart* in England back to the twelfth century when the lords started to lease their demesnes, and sometimes preferred their rent paid in produce rather than cash. This was a commercial arrangement quite different from the petty food rents traditionally paid by manorial tenants. These commercial rents, paid as a proportion of the crop, were frequently associated with ecclesiastical estates and the need to provision large residential households. They also provided a hedge against inflation and allowed the monks to utilize their buildings and take advantage from any upswing in the corn market. In the course of the thirteenth century, the practice subsided as landowners, eager to capitalize more fully on rising prices, resumed the direct cultivation of their demesnes, appointing managers and bailiffs to administer and farm their estates. From the 1370s, in the aftermath of the Black Death, landowners reverted to a leasing policy and sometimes resumed the practice of *champart* rents. Hilton cites several examples from Worcester Cathedral Priory. In 1230 the monks leased the demesnes at Pensax, Worcestershire for the third sheaf, and at Wolverley the lessees rendered half their crop from that part of the demesne that was manured, and a third from that which was not.[11] In this example, the proportion was adjusted to the inputs provided by the lord, underlining the flexibility of the device and the precise calculations that were made. Further west, the Mortimer family in 1347, leased their demesnes at Wigmore, Herefordshire and Radnor, in Wales for the third sheaf.[12]

These examples in the Welsh marches are reinforced by other evidence from that region of sharefarming agreements.[13] These date from the four-teenth century when the lords started to lease their demesnes, and reach a peak during the Welsh raids in the early fifteenth century, as landown-ers in this isolated and unstable area became increasingly desperate for tenants. A common practice was to lease demesne arable annually for sowing in return for money rent or a part of the crop, either the fourth or third sheaf. Kettle describes it as a convenient and profitable way of exploiting demesne lands for which the lord did not have the resources or inclination to farm directly. For example, in the Lordship of Blakemere around Whitchurch, between 1399 to 1403, the lord's share, the third sheaf, of oats from a field previously worth £5 6s 8d as pasturage, was sold for an average of £13 a year. The returns declined after the Welsh raid on Whitchurch in 1410, but it remained in use on the estate until 1468. Similarly, sowing for the fourth sheaf was used in parts of the Lordship of Oswestry. Further south towards Welshpool, in the Lordship of Caus, the parker of Worthen accounted for 7 bushels of mixed corn for land leased for the third sheaf in 1366, and in 1383 the demesne arable at Caus was leased with oxen in return for the third sheaf as a preliminary to leases for years and money rents.[14] We can see that, in this situation, sharecropping was a most useful device for mitigating risk and an effective way of raising revenue from lands which would not otherwise have been cultivated.

This evidence from medieval Shropshire reminds us how historians concentrating on a particular locality and approaching issues from a dif-ferent perspective, can draw out features of peasant life which might elude historians concerned with wider academic debates. Hilton for exam-ple, stressed the exceptional nature of sharecropping or *champart* rent, 'totally submerged by the mass of money rent, labour rent and fixed food rents', to further his arguments on the development of a capitalist economy.[15] To make his case he quoted a list of demesne leases drawn up by Coventry Cathedral Priory in 1411 for 11 of its properties; five were for cash and six for produce rent. Except for one *champart* rent, all of the produce rents were fixed. The fixed rents were paid by individual lessees, while the *champart* – one third of the crop – was paid by customary tenants who took on the lease together. Hilton suspected that this pattern reflected the tenurial relationships between peasants. Likewise, in 1370 the Earl of Warwick leased his demesne at Elmley Castle, Worcestershire, to all the tenants of the manor for the second sheaf, which probably means a half. By 1407 this demesne was leased for cash, reflecting the general trend to money rents in the fifteenth century, and the general preference of landowners.[16]

But why did English landowners prefer money rents to *champart*? This was the nub of Hilton's article. Typically, he addresses the issue in terms of the considerable power the English landlord class exerted over an unfree peasantry, which gave them the freedom to choose the most suitable form of rent, at least until the population collapse of the mid-fourteenth century, when the ratio of land to labour turned in favour of the tenants. Naturally, landowners selected the most advantageous and convenient arrangments. With the proliferation of local markets and the easy disposal of peasant surpluses, money rent became an increasingly feasible option. This was the preferred choice because landowners needed cash to pay taxes and finance expensive lifestyles. Moreover, they were relieved of the burden of collecting their shares, which as Arthur Young described, was notoriously difficult and open to fraud. Hilton's final point is the strength of the English currency. English landowners, unlike their continental counterparts, did not favour payment in agricultural produce as a hedge against inflation or devaluation. In times of rising prices, English landowners could protect their interests by capitalizing on the market and labour services, expanding demesne production and selling their surpluses. The picture Hilton describes is one of well functioning markets, where large landowners could confidently rely on fixed money rents and avoid the bother of *champart*. This interpretation of landlord behaviour has received general acceptance, yet, by Hilton's own admission, it relies heavily on the evidence of large ecclesiastical estates, operating the most advanced systems of estate management in Europe, and even these landowners entered into *champart* agreements throughout the medieval period.

Hilton does not confine his analysis to demesne leasing. The link between *champart* rents and the groups of manorial tenants who leased demesne lands leads him to probe the extent of *champart* amongst the peasantry.[17] He tackles this question through the village land market, but soon encounters difficulties with the documents. As we have seen, manorial court rolls recorded permanent land sales within the village community, but say little about the short-term leasing and subletting, which often lay behind official transactions. Hilton notes that Homans and Harvey both argue that 'the lease at champart was one of the commonest amongst villagers', but their examples were meagre and lacked precision.[18] Homans' single reference from Methwold, Norfolk in 1272, concerning a three year lease of ½ acre does not reveal whether the actual division of the crop was a half or a third. Similarly, Harvey cites just three examples from the Westminster estates, blaming the lack of detail on the oral nature of so many agreements. Thus, problems with the documents leaves Hilton sitting awkwardly on the fence, maintaining that *champart* was of a 'limited and specific nature', but also accepting that it was probably widespread, 'if difficult to document'.[19]

In *The English Peasantry in the Later Middle Ages*, published some years earlier for a different audience, Hilton includes several references to sharefarming that might suggest that his definition of *champart* was, in fact, too narrow and directed to a particular audience. He explains that the actual working of peasant holdings might be rather different from that depicted in manorial rentals. The reality was much fluidity and impermanence through short-term leasing, with peasants continually enlarging and contracting their farms. They did not, however, keep accounts, so it was seldom possible to see the internal workings of holdings. The only insights occurred when confiscations appeared in manorial records. Two prosecutions at Wolverley in north Worcestershire in 1373 indicate the nature of peasant livestock deals, 'for they show disputes over joint trading in sheep and cattle for shared profits'.[20] From 1350 the court rolls recorded litigation between peasants revealing increasing levels of 'inter household traffic'; these included very short-term borrowing or hiring of pasture, herbage, cows in milk, which merged into peasants employing neighbours for contract work, ploughing, mucking and carting. Widows, in particular, made these agreements: in 1375, Emmot of Heye at Newnham, Worcestershire, in the absence of her sons, tried to get a man to cultivate and manure her two acres for her for ⅓ crop as payment.[21] Even by Hilton's exacting standards this amounts to sharecropping.

Hilton's purpose was to play down the existence of sharefarming, but despite his scepticism he provides a clear chronology and structure to the practice. He shows quite clearly that sharefarming in England operated at two distinct levels, as a resort of landowners in times of crises and as a common practice amongst peasants. The examples from *The English Peasantry*, like those from Shropshire, date from the late fourteenth century. This is when sharefarming was most likely to occur as landowners pressed by the shortage of labour and falling corn prices, following the Black Death, sought tenants to lease their demesnes lands. The tenants, once in occupation, needed to make profits, build the capital necessary to farm these holdings and employ labour in the most cost effective ways; hence the recourse to contract labour and agreements with neighbours. Widows were also affected as they competed for labour; hence the greater need for maintenance agreements. At certain times and in particular places, fixed money rents, even in parts of England, were simply not a feasible option for landowners and tenants, or peasants; they needed a flexible device to ease the working of more formal and rigid institutions.

Postan makes a similar point in his work on medieval agriculture and provides, albeit briefly, some of the most illuminating insights into sharefarming in England. With his European background, he understood better than most historians the characteristics and significance of *champart*, and the role it played in medieval village communities. In his *Essays on Medieval Agriculture and General Problems in the Medieval Economy*, he identifies inter-peasant land transactions as one of the most intractable problems facing medieval historians.[22] As Hilton noted, the real difficulty came with the short-term leases, which were seldom recorded in the documents. These were regarded as 'passing events not intended to make irrevocable changes in the owner's property ... current adjustments to the fluctuating circumstances of individuals ... and a mechanism whereby the rigid system of customary virgates could be fitted to the unstable fortunes of peasant families'.[23] He quotes an example from Longleat, where a tenant 'covenanted to have half the profits of 18 acres ... for three years from Thomas Gydye, and also leased 4 acres for three years from another villager, and five acres for four years from yet another man ... altogether, 27 acres'.[24] In this way, they mixed *champart* and money rents to suit their convenience. As to the frequency of *métayage* or *champart* rent, he believed it was more common than we might suppose. 'Leases, *ad campi partem, ad seminandum, ad medietatem* or *pro media vestura* are to be found on all manors cited here'.[25]

There were many reasons why these leases and agreements did not appear in court rolls; some were simply too short to bother with. The Abbot of St. Albans conceded that leases for two years or less could be granted without licence, which would apply for *champart* rents often agreed for a single harvest.[26] Postan compares the English experience of *champart* with the European. Leases abroad were difficult to analyse, as historians relied almost exclusively on monastic cartularies which also tended to ignore short-term leases.[27] The consensus amongst German and French historians was that *champart* was a well-established rent from Carolingian times and predated short-term leases for money. Galassi identifies sharefarming contracts in Tuscany as early as the ninth century.[28] As we have seen, this compares to the origins of land and stock leases in England (Type F) which in its purpose, the provision of credit and working capital to tenants, was closely related to sharefarming.[29]

In *Medieval Economy and Society*, Postan goes further and indicates how sharefarming worked in the village community. He starts by assessing the wealth, well-being and income of the peasant family, comparing it to the experience of the continental *métayer*. He calculates the burden of servile status at 50 per cent of output, once the peasant had paid money rent, labour services and manorial dues, which was not dissimilar to the portion of those that held on share cropping terms *(ad campi partem)* paying half their profits to the lord, 'as in the continental *métayage*'.[30] In fact, the burden could bear more heavily on English peasants, than their French counterparts, as the payments of the English customary tenant were obligatory and fixed, with no concessions made to reflect the harvest or the circumstances of the peasant. The well-being of an English peasant family depended critically on the size of its holding. Contrary to the picture portrayed by manorial surveys, this was not a static entity; peasants were much more able to manipulate and benefit from the land market than their official status would lead us to believe. This applied particularly to smallholders with holdings too small for subsistence, who benefited greatly from the activities of large peasants by supplying them with wage labour, crafts and a range of services. Large peasants, with holdings of three or four virgates, used their surplus to acquire more land. Sources make quite clear that from the thirteenth century, these men were subletting portions, particularly offlying bits and pieces, and dealing in land. Widows, invalids and the old, unable to cultivate their holdings, often resorted to subletting on sharecropping terms. 'Our records abound with references to holdings of this kind, "ploughed" frequently on share cropping terms *(ad campi partem)* by other men usually lessees for a year or two'. The same records contain references to wealthy lessors, leasing widely separated blocks of land often in different villages. The presumption is that land could not have been exploited without recourse to subletting. The large peasant would have no difficulty in managing sharecropping agreements being proficient in the handling and marketing of corn. In a period of high labour costs, such agreements would benefit the lessor. It is possible, Postan suggests, to regard large peasant holdings as 'mini-manors', with portions directly cultivated by the principal tenant, while the rest would bring in rent, goods and services.[31] Wealthy peasants did not confine themselves to the direct purchase of land, they often provided finance and credit for lesser men. This could take the form of lending grain for seed and taking a proportion of the crop in return. There can be little doubt that sharecropping greased the wheels of life in English rural communities. The surprise is how few medieval historians have considered the possibility.[32]

Campbell also makes the case for sharecropping, but from a slightly different perspective. Influenced by the course of Irish history, he likens the role of middlemen, operating in the late eighteenth and early nine-teenth century, to wealthy peasants in medieval English villages. It was they, not landlords, who exploited the manorial system. He differs from Postan and Hilton in arguing that the customary relationship between a landlord and his tenants in practice severely restricted landlord behaviour and created chances for tenants. Protected by the Royal courts, free tenants enjoyed extensive rights over their holdings, paying fixed rents and nominal labour services, while servile tenants held their holdings by inheritance, and certainly by 1300 paid sub-economic rents. In the early fourteenth century, tenants successfully resisted attempts to raise rents in line with prices, insisting on their right to pay no more than a ground rent. Lords anxious not to antagonize them accepted the status quo. In this way, the true beneficiaries of rising prices were tenants paying fixed and customary rents to their lord. They extracted the full commercial value of their holdings by subdivision and subletting, and by using short-term leases and sharecropping arrangements. The effect was to stimulate population growth and rural congestion. Nowhere was this problem more acute than in Norfolk, where sub-tenants scratched a living from the smallest parcels of land, paying exorbitant rents to their tenant landlords which sometimes took the form of sharecropping.[33]

Campbell, like Postan, emphasizes the privileged position occupied by substantial head tenants, who benefited from high prices and low wages, hired labour, produced surpluses for sale, and sent substitutes to perform labour services. They were great assets to their landlords, collecting rents, investing in their holdings, working alongside manorial officials and occupying positions of trust. Lords depended on them to keep the manors and demesnes running, and acquiesced in their commercial activities. At Ramsey Abbey, by maintaining holding size with moderate rents, the monks created the best conditions for tenant investment, attracting men of capital and ability, who nurtured the manorial structure. These 'prudent' lords, managed the tenures and left tenants, skilled in farming, to make the best investment decisions.[34] These decisions included small scale subletting, possibly for a year at a time and on a sharecropping basis, which provided substantial tenants with a means of recruiting labourers to work their holdings and assist in the performance of labour services. Such a strategy was cheaper than deploying family labour to undertake them.[35] In these situations, the land hungry would bid up rents to exorbitant levels, driving up the profits and rents that tenants could charge in a manner not dissimilar to the activities of middlemen in late eighteenth- and early nineteenth-century Ireland.[36] Given these cir-cumstances, it was clear that the standard villein holding of the Hundred

Rolls and manorial extents screened a host of sub-tenants. Even at Ramsey Abbey's manor at Godmanchester, where controls were tight, the chance survival of three subletting lists from the early fourteenth century exposes a plethora of 'short-term, perhaps yearly (one harvest) leases, none of which is recorded in the contemporary court rolls'.[37] And if subletting villein land was hard to police, monitoring freehold land was hopeless. Many freeholders were townsmen who acted as middlemen subletting to peasants. Although least visible, sub-tenants, leasing their land on a variety of short-term agreements, including sharecropping, were probably the most numerous occupiers of land, especially in the crowded parts of east and south-east England.[38]

Once subletting and subdivision had occurred it was very difficult to reverse without provoking violent reaction. Not until population decline broke the deadlock would progress be possible, subdivision halted and then reversed in favour of enlarged farms. The key to this process, Campbell argues, was the gradual replacement of the customary tenures with leasehold, over which much greater control could be exercised. This depended on the elimination of sub-tenancies and the establishment of direct relations between landlords and those who occupied the land. The lax tenurial regime of the medieval period needed to give way to something tighter, with competition displacing custom and landlords adopting a more economic approach in their dealings with tenants. Stimulated by sharply rising prices, this happened from the late sixteenth and early seventeenth century. However, it will become clear that sub-tenants and informal arrangements did not disappear overnight; they continued through the seventeenth and into the eighteenth centuries, often under the aegis of large tenant holdings, just as in the early fourteenth century.

From these accounts a compelling picture emerges of how sharefarming existed and flourished in medieval England. Other historians, notably Razi and Smith, working on the lifecycle of peasant families, offer further insights into how villagers might use sharecropping to exploit their opportunities and resolve their difficulties, and how it might be used to ease a young man into a holding, help him to expand as his family grew, and finally facilitate contraction of the holding and secure maintenance in old age. Only a few historians refer to sharefarming directly, but they all appreciate the significance of informal and flexible agreements, and the misleading impression of village life conveyed by manorial documents.

The inspiration for many of these studies stems from the work of the Russian historian, Chayanov, whose *Theory of Peasant Farming*, has been highly influential in stressing once again the importance of a European perspective.[39] His definition of peasant farming and the family farm is very precise. It is restricted to the lands directly farmed by the family without hired labour. The family farm was shaped by the number of family hands, which influenced the type of farm organization. This meant that the size and nature of the family farm would vary over the lifecycle of the peasant family as it responded to the changing need for land within the family. In a peasant community, this need might be met by renting more land, engaging members of the family in a craft or trade, or by intensifying production, for example into dairying or cheesemaking. At the other end of the cycle, the widow or widower might let land out on short-term leases, or make other informal arrangements. In areas like Russia where the transfer of land amongst peasants was relatively fluid, repartitioning was the norm. In England, peasants leased land, and at a more informal level, farmed to halves.

One of the most authoritative works on the lifecycle of the peasant family is the collection of articles edited by Smith.[40] Only Ravensdale's essay on the position of widows provides a direct reference to farming to halves, but all provide a context where sharefarming may have occurred. From the examples cited by Hilton and Postan, widowhood was clearly a particular stage in the lifecycle when a peasant might have entered into a sharecropping agreement. Ravensdale explains why it was an attractive option. Faced with the prospect of cultivating her own holding, she could either remarry, sell her farm, lease it or employ wage labour. In the land hungry period before the Black Death, a widow's holding was a scarce and valuable asset; lords exploited their position by demanding high fines on remarriage, while widows drove a hard bargain. Maria Buk for example, 'bound her son to take on all the responsibilities of maintaining and working her holding in exchange for a half share in all the crops and the reversion of her land after her death'.[41] After the Black Death, her holding proved a liability, as demand slackened and the cost of labour soared. In both situations, sharecropping provided a solution for her. First, it circumvented the need for sale or remarriage. Secondly, Maria avoided paying a fine for the registering of a lease, reminding us of the advantages of hidden institutions.[42] Finally, she did not have to supervise hired labour as the sharecropping agreement would give the tenant sufficient incentive to perform the task efficiently. Ravensdale makes no comment on the nature of this agreement, nor does he identify it specifically as sharecropping or farming to halves, but clearly, the terms of the agreement, with Maria Buk's son taking full responsibility would be a great relief to her. At the same time it offered inducements to him.

Smith draws our attention to the plight of the propertyless or property deficient.[43] As with Maria Buk's son, it is misleading to think that landless labourers depended entirely on wages. The family cycle and the peasant land market created plenty of opportunities for all, including those without reserves of capital. Sometimes family difficulties could not be resolved within the kinship system. The old, sick and senile often needed the support of neighbours in the high risk phases of the lifecycle. Clark, in her work on social security in medieval England, examined 114 agreements to maintain the elderly for Hindolveston (Norfolk) in 1382, after the Black Death. Of these, three quarters involved no explicit familial tie between pensioner and benefactor. This leads her to reflect that the domestic structure of rural society was 'sufficiently flexible to allow landless peasants a way to find their place by residing with the old, by taking over their holdings in return for the promise of maintenance'.[44] From the example of Maria Buk and her son, it is clear that a system of farming to halves would suit these circumstances nicely.

Sharefarming eased the plight of the old and landless with benefits to both, but how might it facilitate the expansionary phase of the peasant lifecycle, when extra land was needed to provide for large and growing families? Smallholders could increase their wealth by more intensive cropping, short-term leasing, cultivating high value crops, and engaging in by-employments, none of which involved raising capital to purchase land. They could also farm to halves pieces of demesne land, or an old person's holding to set up surplus sons in this way. Once established, they could continue acquiring offlying portions, farming them to halves, rather than employing expensive hired labour, thus further building up the capital of the family. In this way, sharefarming, like short-term leasing, was a flexible method of dealing with phases of shortage and surplus in the lifecycle of the peasant family. Howell's work on Merton College indicates that this may have happened on the manor of Kibworth Harcourt.[45] Here, there was no fragmentation of holdings, and no indication of how dependants and younger sons could be provided for. However, younger sons were left farm equipment and moveables, suggesting they were somehow sustained within a complex extended family group. Sharecropping or farming to halves would offer a solution, where the landholding structure remained static, but the documents remain silent on these issues.

Razi is much more specific in his use and understanding of sharefarming. He cites no less than 21 examples of sharecropping agreements from the Court Rolls of Halesowen between 1270 and 1349, with every indication in the text that this number would have increased after the

Black Death.[46] He argues that the complex economic relationships between peasants, went far beyond the mutual adjustment and co-operation required in an open field system, and included arrangements like sharecropping and the hiring of cows. The Black Death caused a polarization in the village community, with large peasants adding to their holdings by leasing parcels of demesne land, but cohesion was maintained by large peasants offering inducements and employing labour by contract or piece work. This efficient and intensive use of the labour resources of the whole village community enabled rich peasants to tackle the problem of labour shortage much more successfully than the lord. Court rolls showed tenants employing servants and wage labour, but increasingly they engaged other big, middling and small landholders to do contract work for them. Razi cites the example of Richard Moulowe employing almost everybody in the parish, and letting various plots of land under sharecropping agreements. In 1375, he sued Henry Townhall, another *kulak* to whom he leased plots of lands for half the crops, which he had taken for himself.[47]

In this way, large farmers or *kulaks* became heavily reliant on piece work, contract labour and sharecropping. It was tricky to administer but much more efficient in a period of high wages than using direct labour. The crucial point, as Postan and Campbell also make clear, was their ability to monitor, control and exploit such a system. Although the gap between *kulaks* and villagers widened after the plague, they had a vested interest in the survival of the tightly knit village community based on reciprocity and co-operation. It was this mutual self interest that made these systems work. In contrast to the *kulaks* who were able to work with the villagers, the Abbots of Halesowen were increasingly isolated. Their situation worsened after the Black Death, as the balance of power shifted from landlords to manorial tenants. Thus, we can see that the fall in corn prices and the shortage of labour created imperfections in two key markets. First, it forced landowners to lease their demesnes as best they could. This might mean having to supply the seed and sharing the crop with the lessee, as they were not in a position to demand fixed money rents. Similarly, large tenants had to offer inducements to husbandmen to work the land, which might take the form of sharecropping. It was these sort of 'imperfect conditions' that explain the existence of sharefarming in English rural communities in the late medieval period.

The remainder of this chapter will examine more fully the variations of sharefarming as they existed in the medieval period in terms of our taxonomy. The most common were the stock and land lease (Type F), cow leasing, (Type G) and corn rents (Type H). The shared ownership of crops and animals (Type E) no doubt occurred in this period, but specific evidence has only been found in probate inventories. The study does not extend to Scotland, Ireland and Wales as the tenurial traditions in these countries differ quite significantly from England. However, a discussion of these Celtic variations, principally steelbow tenure, illustrates the extent and complexity of the practice in all parts of the British Isles (Type A).

The earliest variant, as we have seen, was the stock and land lease. In his pioneering study, *Rural England*, Lennard notes that the farming out of manors on stock and land leases was 'so common that this mode of estate management, inherited from Anglo-Saxon England, must at least rank in importance with the system of tenure by knight service introduced by the Normans'. In this arrangement the lord provided working capital for tenants in the form of stock, cows, draught horses, oxen, sheep and pigs. He cites scores of examples from Norfolk and Suffolk on several ecclesiastical estates. In one case, the Canons of St. Paul's stipulated that on the death of Mabel, widow of Galium, half the stock should return to them, indicating that sharefarming co-existed with the stock and land lease.[48] Thorold Rogers, in his description of the origins of capitalist cultivation, placed much emphasis on the use of land and stock leases, by the lord when he let his demesnes land to farm.[49] These leases usually involved the leasing of dairy cows at 5s or 6s 8d a year and sheep at 1s, with the tenant agreeing to restore them in the same condition at the end of the year. The bargain implied men with the power of contracting, far removed from the picture of the landless labourer with no rights or means. Thorold Rogers believed cow leasing to be 'exceedingly common in the middle ages, so common it was the rule on many estates' . The practice 'endured longer than any custom handed down from the middle ages, for it is constantly referred to by Arthur Young'.[50]

Stock and land leases were redefined in the late fourteenth century, when landlords abandoned direct cultivation and leased their demesnes. The new leases were generally for short terms of seven to ten years, with the lord covenanting to repair buildings and share the risks of disease. The lord insured against murrain in sheep while the tenant took responsibility for scab, considered curable by proper husbandry and appropriate remedies. Besides this, the lord assisted tenants when prices of corn were low, or when they suffered losses, 'the lord learnt early that he had to submit to part of the casualties which the tenant endured'.[51] At this stage, the arrangement benefited both parties: it met the tenant's need for working capital and gave the lord the option to resume farming if he wished. However, from the mid-fifteenth century, as the economic climate shifted firmly in favour of the tenant, allowing him to exploit the system, drive the hardest bargain, buy the cheapest stock and exaggerate his losses, the stock and land lease petered out as landlords wound down and simplified their estate management.

Harvey in her work on the Westminster Abbey Estates refers to stock and land leases and makes the connection with *champarty* rents, confirming the relationship between the two forms of tenure. She identifies stock and land leases as a feature of the later middle ages, as the Abbot started to lease his demesnes.[52] He provided lessees with the necessities of farming – livestock, seedcorn, poultry, draught animals and implements – to get them started.[53] This method had the advantage of preserving the basic structure of the holding, the equipment and stock, should the Abbot wish to take the demesne back in hand. The practice started to die out from the 1420s as this prospect became remote. The result was a sharp fall in rents. For example, the rent of £15 for 400 acres of arable at Islip was halved when the stock and land lease ended.[54] In other words, the monks had received an income from their existing stock and equipment, but did not seek to renew it. At the same time they became more tolerant of subletting. As Harvey says, the lessee chosen by the Abbot, probably acted as much as a middleman as a cultivator; the reward being not the right to occupy a nice farm, but the right to dispose of it at a profit. Leases contained the usual restrictions on subletting, but that did not mean it did not happen, just that permission was required. In her view, there were no grounds for assuming extreme fragmentation, but plenty of opportunity existed for informal agreements between peasants, particularly as the monks rarely allowed leases for other than short terms. For leases of free land, indentures were the preferred form, but for customary land, verbal agreements were the norm. Consequently, in these cases, when agreements were made orally, we know nothing about their terms. It was in this context that Harvey first mentions '*champarty*' leases, as being 'probably very common'.[55] 'Short-term leases and *champart* were of a piece with the precarious nature of peasant existence and probably remained the typical arrangement until the end of the middle ages'. Although arrangements for 'crop sharing' became rare in the fourteenth century, both the abbot and convent required some of their lessees to pay the whole or a proportion of their annual due in corn. For example, John Clerk, who had a lease of a manor at Kelvedon in the 1350s, owed 20 quarters of wheat per annum, besides a money rent of £13 6s 8d. In the fifteenth century, the Abbot switched to money rents, but the convent continued with corn rents until the sixteenth century. Of interest, is the fact that Harvey describes this type of rent as sharecropping. Strictly speaking these agreements involved a fixed payment in kind, rather than a percentage of the total product. This confusion highlights the fluidity surrounding these types of arrangements.

Corn rents, of the type described by Harvey, are a particularly interesting variant of sharefarming. Britnell, in his survey of medieval Britain, emphasizes the role of produce rent in a commercializing economy.[56] However, as he explains, our understanding of this process is hampered by the changing nature of estate records in the fifteenth century. The quality of the material declines sharply. For while historians enjoy a profusion of manorial accounts and court rolls from the thirteenth and fourteenth centuries, showing how demesnes were stocked, cultivated and managed, they petered out in the fifteenth century when demesnes were leased, so it is rarely possible to know how the land was actually used. When detail of farming activity does survive all kinds of leasing arrangements emerge. This is why the Paston letters are so valuable. From the 1440s to the 1470s, we find the family heavily reliant on barley rents. Although Britnell does not make the connection, corn rents in Norfolk were closely linked to sharecropping, or sowing to halves as it was often known in that county.[57] In fact, a reference to the Pastons sowing to halves at Edingthorpe in 1420 was found in the documents.[58]

Corn rents are not normally associated with highly commercialized and advanced agricultural systems, but they were common in fifteenth-century Norfolk and survived until the eighteenth century.[59] They appear to have been a legacy of demesne farming in the thirteenth and early fourteenth centuries. As we saw in Hilton's article, when lords started to lease their demesne, many preferred to receive rents paid in barley. This policy made sense, particularly in a grain-producing area like Norfolk. No region could export grain and malt so readily, therefore it paid Norfolk landowners to stay in the market and exploit the bulk trade in corn. Moreover, they were already geared up for the trade, possessing storage facilities, established contacts and the marketing skills of their estate officials. Barley rents were levied at a fixed amount, often two bushels per acre, so landowners enjoyed certainty. They also benefited from any rise in prices, and were protected when the harvest was deficient. In times of oversupply and low prices the barley could be malted. The advantage over sharecropping was the assured return and the ease of collection and management. The Pastons were especially well placed in the barley trade. Their estates were concentrated principally in north east Norfolk, an area of rich soils, intensive cultivation and high yields. Most importantly, their estates enjoyed easy access to London and European markets through Great Yarmouth.[60] It is evident from their letters that barley rents were a regular feature of the Pastons' income. In 1473, John Paston II beseeched his steward, Richard Calle to 'make a good bargeyn for my ferme barley in Flegg, so that I might haue money now at my beying in Ingelond'.[61]

Barley rents were not confined to north east Norfolk. They were common near the coast, particularly on ecclesiastical estates where they were associated with the payment of tithes. Britnell cites several examples of these estates collecting corn rents and tithes from their coastal properties, including the Ramsey Abbey manors at Brancaster and Burnham Deepdale, close to the port of Wells in north Norfolk.[62] Further west, and a few miles north of Kings Lynn, the Dean and Chapter of Norwich also collected barley rents from their manor at Sedgeford.[63] Unlike east Norfolk, this was an area of light soils and extensive agriculture. The lands of Sedgeford were organized around a classic infield and outfield system, with the fertile river valley in permanent cultivation and the sandy *brecks* ploughed intermittently. About 50 per cent of the *brecks* were cultivated at any one time, with wintercorn, barley, barley with peas or oats, before being fallowed and grazed by sheep for a number of years. The system was mutually beneficial, with sheep raising fertility of the soils, and cultivation improving the quality of the grass, but it involved quite intricate organization. The lord enjoyed grazing rights, or rights of foldcourse, over the open fields and *brecks*, which required much collaboration and negotiation with tenants; when the *brecks* were out of cultivation the tenants had to be compensated with land elsewhere. This integrated system, with lords engaged in sheep farming and large tenants cultivating the arable, lent itself to produce rent. The tenants stored the grain in huge barns and supervised sales through nearby Heacham and Kings Lynn. In this way, an 'archaic' rent was found to be the most profitable option in the most and least fertile parts of Norfolk, driven by a well organized corn market.

The popularity of barley rents may explain the origins and appearance of letting to halves, principally for corn, on well known Norfolk estates in the seventeenth century, for it is only a short step from paying a fixed amount in barley, to paying a fixed share, but here the problems with documentation are acute. Barley rents were a fixed, well-established, long-term arrangement, while sharecropping was essentially a short-term expedient and often based on oral agreements. As Britnell explains, the problem is compounded by the deterioration of estate and manorial records in the fifteenth century, which limits the amount of information on farming practices and agreements at that time. However, a quirk in the accounting system at Raynham shows that sharecropping did occur in this context.

For a short period from 1470 to 1485, the estate accounts of the Townshends of Raynham include details of stock and grain farming on the back of the account, which show that the family entered into sharecropping agreements with the tenants of their demesne lands at Stibbard and Little Ryburgh.[64] Roger Townshend, a renowned lawyer and enterprising landowner kept huge sheep flocks, like the Pastons and Le Stranges, and sold great quantities of grain. Sometimes, he resorted to sharecropping to cultivate the arable, providing his tenants with seed to plant, and in return for growing and tending the crop, they took half of what was produced. By 1485 at least four tenants were sharecropping, each growing crops of barley, oats, wheat and maslin for Roger Townshend. Some tenants at Raynham stored large quantities of grain, but in this instance, sharecropping did not occur; it was in effect a loan, with the landowner providing the working capital and the tenant repaying him in grain at the end of the year. These accounts, for this very short period, reveal a variety of informal methods used to cultivate grain. Sharecropping stood somewhere between direct farming, letting the land for cash, and barley rents. The stock accounts reveal another service performed by the tenants, the threshing of grain for which they received a part of the crop for their labours. Unfortunately, after the changes to the accounting system in 1485 all detail of farming activity ceases, so we have no idea whether sharecropping continued at Raynham before it reappears in the seventeenth century. However, as the arrangements of the fifteenth century were not far removed from the letting to halves agreements introduced at Raynham in the 1690s, we might assume that the practice had continued in the intervening period.

This hunch is confirmed by a chance reference which gives a further explanation as to why Norfolk landowners resorted to barley rents and sharecropping agreements. A petition to Mary in 1553, from the common people of Norfolk, shows them complaining about landowners switching from charging money rents of 6d, 7d and 8d per acre, to barley rents of three to four bushels to the acre, or 'elles the therde sheffe'.[65] At that time, as Hoyle points out, barley prices averaged 8s 6d per quarter, which meant rents of 4s were being demanded for lands previously rented at 1s per acre. Thus, sharecropping and barley rents were an extraordinarily effective way of raising rents in line with inflation and of exploiting labour as wages also started to rise. As we shall see in the next chapter, for reasons such as these, the French agriculturist, Olivier de Serres recommended the practice in the early seventeenth century.

Britnell's survey extends to most parts of medieval Britain and includes a strong Scottish and Irish perspective. This is useful as corn rents existed here from the earliest times, accompanied by various forms of share-farming. As we have seen, in his work on the Pastons, Britnell emphasizes the role of produce rent in a commercializing economy. His first Celtic examples come from Bute and Arran where the rents to the Scottish Crown in 1496 included a third paid in grain and stock. It is not clear how this arrangement worked, whether the stock and grain had to conform to a fixed monetary value, or whether the landowner was prepared to accept a shortfall when prices were low. His example from Gaelic Ireland is more clear cut, where rents were also paid in grain. In the sixteenth century and earlier, it was normal for even quite substantial tenants to hold their land on a sharecropping scheme, by which they supplied the seed and equipment, and paid a quarter of the crop in rent.[66] What is interesting is that the tenant paid only a quarter of the crop rather than the more usual half, as at Stibbard in Norfolk. As we saw with Hilton's example in Worcestershire, the proportion of the corn depended on the inputs provided by the lord, and the value of the tenant's contribution. So, it was a precisely judged arrangement.[67] In isolated areas, larger tenants with capital would be able to negotiate advantageous agreements. In Ireland, as in England, the balance of power between the landowner and tenant had, by this time, shifted firmly in favour of the tenant, further strengthening his bargaining position. In Norfolk the position of the landowner was eased by the existence of well organized grain and malt markets. In both these situations, remote Gaelic Ireland and accessible, highly commercialized Norfolk, forms of sharefarming were used, in the first case, to compensate for deficient markets, and in the second to facilitate the exploitation of a particularly well developed commodity trade. Similar comparisons can be found in Scotland.

Dodgshon provides a detailed explanation of how sharefarming worked in medieval and early modern Scotland.[68] Steelbow tenure, the Scottish term for sharefarming, was associated with the leasing out of demesnes from the late thirteenth century, and as the term implies has its origins in military tenure. It most closely resembles the stock and land lease in that landowners provided the tenant with stock, seed and equipment as part of his lease, on condition that the tenant rendered up the same quantity and value at the expiration of his term.[69] In some cases, the tenant paid his rent in kind either as a fixed corn rent, or as a share of the produce. Like farming to halves and corn rents, steelbow tenure played a significant role in the development of agrarian capitalism, allowing landowners to profit from the growing market in grain and enabling tenants to build up capital and equip large farms. As farm sizes grew, steelbow tenure enabled the ambitious tenant to reach well beyond his immediate circumstances and farm really large holdings. The practice continued in Scotland until the mid-nineteenth century.

Dodgshon describes how steelbow worked in eighteenth-century Teviotdale, where it went under the name of 'thirders and teinders', and how it could descend to the level of farm servants and sub-tenants.[70] In 1742, for a farm in Roxburghshire, the tenant entered into an agreement with his farm manager, William Legetwood, who was obliged 'to sow 40 bolls of oats, beans and peas, whereof he is to have a third and teind shorn and set up'.[71] In other words, the farm manager paid the tenant a third of the entire crop, with the teinds (tithes) owed by the farm. Legetwood had also to answer for all carriages and daywork owed by the farm to the landowner. As regards pasture, he was allowed only a third of the haycrop in return for mowing and leading the entire haycrop to the stackyard, and enjoyed the right to graze a number of sheep in return for managing the tenant's stock on the farm. From this agreement, an essentially medieval arrangement, involving the payment of rents in kind and tithes, and the performance of labour services, was adapted to serve the needs of a capitalist market. As in France, as Dodgshon makes clear, produce rent and sharefarming could prove more exploitative than money rent as landowners tried to capitalize on rising grain prices and other feudal residues such as forcing tenants to use the lord's mill. The combined effect of these factors was, as in France, to reduce the economy of many lesser tenants to one of severe risk as regards its sufficiency. There was an old saying, 'one [sheaf] to eat, one to sow and one to pay the laird withaw'.[72] Those on the lowest rungs of the ladder had little surplus beyond the three basics: rent, seed and subsistence. Stock losses could also be punitive: 'when the muck runs out the barley seed is done for', leaving them only enough to eat and pay the rent. Between 1695 and 1703, bad harvests led to a loss of a third of the Scottish population, emphasizing the need for subsistence crops, like oats. However, with the union in 1707 and access to the English markets, prosperity grew and generated a movement to money rents, but steelbow was still a significant feature of Lowland farming in the late eighteenth century as Adam Smith pointed out.[73]

Steelbow tenure was also associated with the more paternalistic approach prevalent in the Highlands and the west of Scotland. Many peasants, as Smout describes, 'held their land by steelbow tenure by which the landlord provided stock and seed with the land in the same way as the *métayer* system in France'.[74] In this area, as in Gaelic Ireland, the landowner, had little option but to help the tenants, by providing working capital and sharing the risk of farming. These remote and geographically difficult areas on the Atlantic fringes of North West Europe presented severe problems for farmers, forcing them to adopt risk aversion strategies. This kind of extreme environment guaranteed the survival of steelbow tenure in these areas until the nineteenth century.[75] From these examples in the remote Highlands and improving Lowlands of Scotland, we can see that steelbow tenure could be used in a variety of ways in response to geographical, economic and social conditions. Its principal attraction was its flexibility and adaptability to different situations. The downside was the bother of collecting the shares. In this context, the role of the 'tacksman' in Scotland and the middleman in Ireland, who held blocks of land from the laird and leased out parcels to sub-tenants, is significant.[76] It was at this level that sharecropping agreements could be most readily enforced, but when documentation becomes extremely difficult. As in England, estate papers in Scotland and Ireland conceal the extent of subletting being only concerned with head tenants. There was no reason for the parties to keep records, in fact they had every reason to keep their arrangement as covert as possible, lest a landlord or his agent discover just how much the direct tenant's property was worth and raise the primary rent.

Another feature of medieval farming communities in Scotland that attracts comment in this context, is that they continued to be organized in a more collaborative, risk-sharing way than English villages.[77] The common unit of settlement was the farmtoun. There might be several farmtouns in a parish, which centred on a notional farm, being an area that a ploughteam could keep cultivated. The farms could be worked in different ways. A single husbandman might enjoy his tenancy as a whole and cultivate it independently with the help of lesser men, his sub-tenants and servants. It was frequent for two or three husbandmen, or as many as 16 to 18, to share a farm as joint or multiple tenants, which could involve a variety of arrangements. At its most basic level, husbandmen held land in common, cultivated land in common, shared the crop communally and paid rent as a lump sum. This type of mass tenure was practised in Ross and Cromarty and the Hebrides into the eighteenth century, but there was no evidence of it elsewhere by the sixteenth or seventeenth centuries. Slightly more sophisticated was the runrig system, which consisted of separate holdings of scattered strips to ensure a fair distribution of good and indifferent land, as in the English open field system. Under periodic runrig, strips were reallocated at intervals; sometimes strips were consolidated into blocks, known as rundale, and intermixed with lands of joint tenants. In this context, we would expect to find examples of partnerships and half shares (Type E). Finally, the stage was reached when it was possible to divide the arable into complete separate, consolidated and independent holdings.

These systems appear bizarre and inefficient, but like the open field system in England, they were designed to support a population and not primarily to supply the market. The long survival of such arrangements in Scotland was a reflection of extreme conditions and the need for risk aversion strategies to insulate the inhabitants from the full impact of the climate, poor soils and deficient markets. The situation was partially relieved by access to English markets from 1707. However, the full benefits were not realized until political stability and better communications were achieved in the second half of the eighteenth century. It is from this perspective that Adam Smith's implicit criticism of backward Scottish farmers with their steelbow tenure, in contrast to their progressive English neighbours should be viewed. For the English such practices were seemingly a distant memory, but not so distant as Smith and his contemporaries believed.

Although historians are severely hampered by the nature of the records, it has been possible to construct an outline history of sharefarming that existed in England before 1500. It was associated primarily with periods when landlords leased their demesnes and were unable to secure money rents, or, for some reason, preferred part of their rents to be paid in produce. At the same time flexible short-term arrangements, including sharecropping, existed between peasants, frequently under the cover of subletting. Rarely documented, these habitual practices eased the village community through the vicissitudes of life and enabled peasants to exploit their opportunities. It is not possible to quantify the extent of sharecropping in medieval England, as discretion between the parties remained an essential characteristic.

When the search is extended beyond England it becomes clear that forms of sharefarming were integral to the organization of agriculture in the more remote and unfavourable parts of the British Isles. In these environments inhabitants needed risk aversion strategies to protect themselves from the impact of very difficult markets. These were not fully addressed until the late eighteenth century and, in the case of Ireland, not until the late nineteenth century. The Scottish examples illustrate how sharefarming was used at opposite ends of the economic and social spectrum. In the Highlands and Islands, deplorable conditions required landowners to behave paternalistically and look after their tenants and crofters. While in the improving Lowlands, it enabled land-owners to exploit an increasing profitable grain market. In this way, steelbow provided security to impoverished rural communities and at the same time formed a bridge to a fully capitalist agriculture, anticipating a time when tenants would confidently lease large farms for terms of years on fixed money rents, supplying their own working capital. In England, a similar contrast could be found in the troubled Welsh border counties of the fifteenth century and the increasingly prosperous corn fields of Norfolk.

Hilton, for his own purposes, plays down the significance of *champart* rent, yet recognition of the practice was found in the accounts of leading medieval historians. References were brief, but significant. Postan and Razi, with their European backgrounds, both understand fully its place in medieval rural society, while Campbell has arrived at this position through his experience in Ireland. These historians disagree on many issues, but all appreciate that sharefarming played a role in English village life. Yet none of them, apart from Hilton, has focused on the practice or considered its existence in a wider context. In European circles, Hilton's paper effectively closed the discussion, while in England it seems to have extinguished what limited interest there was in the practice, that is until Campbell's contribution published in 2005.[78] The difficulty remains the lack of documentation, which becomes progressively more difficult in the late fourteenth and fifteenth century, as landowners leased their demesnes and when sharefarming was most likely to occur. However, when problems with the documents ease in the late sixteenth and seventeenth centuries, more evidence of sharefarming comes to light.

4
Sharefarming Comes to Light: Early Modern Evidence

Much more evidence of sharefarming comes to light in the early modern period. Although, as we have seen, sharefarming was not exclusively a seventeenth-century phenomenon, there were clearly reasons why sharefarming becomes more visible during this period, and why it may have been more prevalent. New evidence is partly associated with improved documentation of all kinds and better government administration. For example, by the sixteenth century, reneging on a farming to halves agreement in parts of Wales had become a criminal offence, punishable in a court of law. Suggett found scores of cases in legal records in Radnorshire, indicating that the practice in this region was so widespread that it needed to be institutionalized and framed in law.[1] In England, from the 1550s, probate inventories record the goods and chattels of deceased persons who sometimes left 'moieties' of crops and livestock, with references to 'the partible way', 'halfendeale' agreements and animals 'at half crease' (types A, B and E in our taxonomy).[2] From the early seventeenth-century estate records also began to improve significantly, as landowners, responding to the price rise of the late sixteenth century, sought to modernize the management of their estates. At a lower social level, greater literacy played a part, as yeomen and minor gentry began to keep diaries and account books, showing a concern for profit and loss and an awareness of the new possibilities. At the same time, agricultural manuals appeared in increasing numbers, explaining to landowners and farmers how they might achieve their objectives.[3] This chapter will show that types of sharefarming existed and developed across much of England during the sixteenth and seventeenth centuries. Explaining why this happened, and why we know more about it, will set the scene for the Norfolk case studies that follow in Chapter 5.

Although some historians are familiar with the concept of sharefarming in the early modern period it rarely features in their accounts. An exception is Kerridge who highlights the way that manorial records, particularly manorial surveys, conceal layers of sub-tenants, share-tenancies and sharecropping. By the sixteenth century, subletting by lessees of demesne lands, freeholders, copyholders, and even sub-tenants, had reached such proportions that it had created a 'tenurial illusion'.[4] Harrison illustrates this point precisely in his comparison of two sixteenth-century manorial surveys of Cannock Chase.[5] The first survey dating from 1570 was a conventional survey of tenancies, while the other from 1554 recorded, atypically, not only tenancies but sub-tenancies. The latter showed that 64 per cent of the land was sub-let. Sub-tenants included not only landless labourers, but existing tenants, seeking to expand their holdings who paid their rents in both money and kind. These arrangements may well have involved sharecropping, but unlike Kerridge, Harrison makes no mention of the practice.

In his discussion of share-tenancy and sharecropping, Kerridge explains their exclusion from manorial records, 'implicitly because they did not constitute an estate in land or real chattels, and sometimes explicitly'. For example, a custumal of 1639 forbad the underletting of copyholds for more than a year and a day without a licence, 'except it bee to thirds and halfes'. Thus, as manor courts disregarded letting to halves and thirds, the evidence is slight. Nevertheless, he claims the practice extended to 'the Cotswolds, Chiltern and Oxford Heights, Fen and Cheshire Cheese, the Vale of Evesham, the Wealden Vales, the Vales of Hereford, the Midland and Lancashire Plains – in a word where much cultivation was undertaken; and dairy herds and pastures were sometimes let out in a similar fashion'. But evidence remains a problem. Agreements for sowing to halves and thirds were usually only by word of mouth and records of them were purely fortuitous; copyholders frequently let their lands to thirds and halves, but just how often remains unknown.[6] This view accords with our findings, but Kerridge remains the sole example of an historian making such claims for sharefarming in England.

Significantly, Kerridge's more specific examples of farming to halves date from the early seventeenth century. He reminds us that Robert Loder let the demesne farm at Harwell Manor to halves. An entry on the first page of his farm account notes that when he came into his estate in 1610, lands which had been his mother's jointure, and which were in his uncle's use, he had 'put forth to halfes'. For that reason, Loder tells us, 'I know not, what or how much seed there was sown upon it'. In 1612 he compared the return on tilling the land himself and letting it to halves: 'What I have gotten by tilling my lande my self more then I should have done yf I had put it to halfes. Memorandum that yf I had put my lande to halfes it would have cost me half of the proffites aforsayd that is to say, half [£290 6s 1½ d] which [£145 3s ¾ d] soe that my charges being … £104 16s I have gotten by tilling my land myselfe, the residue of the £145 3s ¾ which is £40 7s ¾'. By not having to surrender half the profits, he calculated that direct farming was more profitable than letting to halves. By this type of close management, Loder increased his income from £150 to £200.[7] Loder also compared the cost of keeping servants with 'putting forth my lande to tillage' which might refer to contract farming at flat rates, rather than letting to halves. In this way, Loder kept all his profits. Once he was in residence and could supervise wage labour with little cost to himself, it made sense to follow this course, rather than to let his lands to halves. But in his absence, when he was reliant on others, the reverse was true.[8]

Kerridge also describes how sharefarming agreements were used by landowners and graziers to implement his 'up and down husbandry'. Graziers frequently leased closes singly to farmers for a term of years for ploughing to halves or thirds.[9] This gave graziers the advantage of temporary leys without the bother of cultivation. The period also saw the rise of itinerant woadmen and flaxmen. Woad was invariably grown as the first crop of newly broken pasture. By special agreements the contractors took a single industrial crop and a few subsequent corn crops and then passed the land on to someone else. This is similar to 'taking a crop', a practice still used in the Fens and East Anglia often to grow catch crops of vegetables and roots. The advantage of these arrangements was flexibility; they enabled landowners, farmers and contractors to experiment with new crops and enterprises, while minimizing costs and spreading the risk. Significantly, these developments occurred when farmers were starting to make precise calculations as to the profitability of different courses of action. In certain situations, sharefarming offered the most efficient solution.

References to sharefarming on the chalk downs can also be detected in Bettey's work on the floated water meadows of Wessex. He mentions, in passing, that tenants shared the cost of construction, shared the rights to watercourses, and quotes an agreement in which they shared the profits with the landowner. In 1635, at Winfrith Newburgh, the Earl of Suffolk agreed with his tenants to improve Winfrith Mead, 'wee the freeholders, leaseholders and copyholders doe hereby for us and our successors covenant and promise to and with the said Earle, his heirs and assignes, the value of one Moiety [half] of the profits that shalbe raised by meanes of the said improvement to be valued by two different persons'.[10] In this way they shared the cost and the benefits. The Earl proceeded to construct watermeadows in several manors along the Frome and its tributaries, altering radically its appearance. Bettey, however, makes no comment on this arrangement, but we can see that it ensured the co-operation of the tenantry in an expensive and highly risky venture. Chapter 5 will show how the Le Stranges entered into similar agreements with their tenants when they drained the marshland at Hunstanton, Heacham and Holme.

The association with high risk and costly enterprises can also be seen in the Welsh Marches in the sixteenth century. Suggett explicitly links 'sharecropping' to the growth of the cattle economy in that region and the construction of the longhouse. Instability and untenanted holdings in the late fourteenth and early fifteenth century had forced landholders to revert to pasture and more extensive farming regimes, and in that process, they resorted to 'sharecropping'.[11] In the sixteenth century, benefiting from political union and the rising demand for food in England, hill farmers turned to breeding and fattening livestock. Suggett describes how they used sharecropping as a way of accumulating stock and capital. Typically, they hired cattle and sheep from the lowland areas, Shropshire, Herefordshire, and east Radnorshire and kept them from year to year for 'half the increase' – that is half the lambs or calves, half the cheese or half the profit on the fattened cattle.[12] The key to profitability was maximizing the use of the upland commons. Stock spent the whole period from spring to autumn on the commons, and then a proportion returned to the farmstead to be over-wintered on hay. Hence the development of the longhouse, used to protect the cattle from theft, exposure and starvation in the dark winter months.

The practice has come to light as these Welsh farmers, in contrast to their English neighbours, entered into written agreements, supported by a bond of about £40.[13] If the conditions were broken, the penalty could be recovered by an action for debt in the Court of Great Sessions. The result is that numerous disputed profit-sharing agreements can be traced in the legal records, revealing both the widespread nature of the practice and also the problems encountered with it. The loss of animals when on hire was a serious matter. The general rule was that if animals died from sickness – 'murren', 'rots' or 'mountain evil' – both parties made up the stock jointly. But if any died through negligence, or were stolen, then the sharecropper was solely responsible for making up the numbers. Many owners of cattle sued for the penalty when the profits were not delivered to them or their stock was not returned on time. Usually the agreements involved between six and ten kine and 20 to 40 sheep, although larger numbers of stock could be hired for the summer months. In one of the earliest, in 1544, Thomas Morgan of Arkstone, Herefordshire delivered 20 kine and 20 sheep to Meredith ap Howell of Cwmteuddwr for a period of six years at £6 per year. Occasionally, tenements and stock were leased together. In 1549, a tenement in Llanstephan, with its stock of 20 cows and 40 ewes with lambs, was leased for three years at £6 per year. In 1581, a tenement in Bleddfa, with 12 cows and 120 wethers to be reared for mutton, was leased for three years for half the increase. In Llanbister in 1559, 24 cows, six oxen, 12 score sheep and five sows were delivered to David ap Cadogan Vayne for three years on a sharecropping basis.[14]

In terms of our taxomony, it is clear that Suggett fails to distinguish between variations of the practice, and his use of the term 'sharecropping' in this context is misleading. There were in fact several methods of hiring out livestock: placing ewes at halves where the lambs and wool were divided (Type A); grazing cattle for half the increase, or profit, on the fattened stock (Type B); leasing out stock with a tenement in a traditional stock and land lease (Type F); leasing cows in return for a money rent or half the cheese or butter (Type G); or finally, simply paying a headage rate for grazing stock, known as 'at tack'.

Suggett includes a case study to show how sharefarming worked on one particular farm, Nannerth Ganol situated in the upper Wye valley, near Rhayader, part of the monastic grange of the Cistercian Abbey of Strata Florida. About 1520, the abbot granted a moiety [half] of Nannerth on a long lease of 99 years to Bedo ap Steven who paid 5s yearly, with a further 3s 4d for the moiety of the summerhouse and its pastures. The other halves were leased separately. The longhouse was built from the profits of the lease and the successful marriage of his son in 1555. Litigation at the Court of Great Sessions provides illuminating details of the activities of his sons, Thomas and Edward, with nearly 20 actions for debt in the plea rolls involving livestock. These were often sharecropping agreements, aimed to maximize the profits from the vast pastures. In 1556 Tho Bedo agreed with Roger Tyler of Presteigne to graze a stock of 153 ewes, and to deliver the skins of those that died. Some years later Roger Tyler junior was seeking the return of 20 cattle and 40 ewes, leased to Thomas for three years for half the increase of calves, lambs and cheese. Later in 1561 Thomas and John Fletcher of Herefordshire delivered to Edward ap Bedo 18 bearing mares to maintain from year to year, in return for half the colts. Edward was to make up the stock if any were stolen. Unusually, some of the writs against the Nannerth brothers were pursued to judgement and they spent periods in jail. In 1557 they were found guilty of cattle theft – longhouses could be used to hide cattle as well as protect them.[15] The impression gained from these records is of rural entrepreneurs determined to exploit fully the resources available to them, wheeling and dealing in livestock, but sometimes falling foul of the law.

The trade in fat cattle prospered in the sixteenth century, but in the seventeenth century it declined, as the keeping of prime cattle gave way to stores, with the profit taken by those who did the fattening on the lush pastures of Shropshire and Herefordshire. By the eighteenth century, sheep were the dominant feature of the pastoral economy and apparently sharecropping disappeared. We can see that during a critical period of transition it had allowed under-resourced farmers in this inhospitable region to take advantage of a new market for livestock. The longhouse and the legal records survive as a legacy of the strategy employed by farmers at this time. Farming at halves re-appeared in the early twentieth century, as the railway increased accessibility to this remote area and encouraged hill farmers to build up their livestock enterprises, but whether it survived during the intervening period remains a conundrum.[16]

The question can be partially answered through probate inventories. A sample of Shropshire inventories, transcribed by Edwards and dating from 1600–1707, are evenly balanced between livestock and corn, as most derive from the North Shropshire Plain. Two inventories of father and son, both shoemakers from Wellington, indicate a full partnership in 1639, while four references state that corn was sown to halves or parts; another shows that a husbandman owned half the lambs and wool, a sign that he held ewes at halves.[17] There were also references to third and quarter parts. The evidence of farming to halves is limited, but sufficient, especially given the experience across the border in Radnorshire, to suggest that the moieties represented forms of sharefarming.[18]

We have also been able to draw on inventories from three other counties; Norfolk, Cornwall and Kent. Overton transcribed several thousand inventories for Norfolk and Suffolk but only extracted quantitative information and did not transcribe any other material systematically.[19] Despite this we have a few examples of Norfolk inventories which refer to sharefarming, including George Belowe, husbandman of Postwick, in 1587, who was in possession of 'half of two akers of Ry ptable between Richard Funnel and George Belowe xviii, halfe of six akers & ½ of barly ptable between the said Funnel and the said Belowe 6s.' In 1633, Peter

Turner, yeoman of Wreningham left 'an halfe of ii acres of ptable wheat in the ground with Richard Ryddall'. More typically, in 1628, Robert Underwood, yeoman of Deopham, left '6 cows, one ptable bull ye half 10s ... One old mare with foale ptable at 20s = 10s ... Seventeen acres of winter corne ptable at 30s, an acre to be pted by the bushel and half the grass upon the borders ... ii acres & a half of pease ptable & 10 acres of barlie to be sowen ptable with the seede at 20s p acre.'[20] Rather than 'to halves' the term used was 'ptable', or 'partible', the same as Sir Hamon Le Strange used at Hunstanton.

The samples of probate inventories from Cornwall and Kent were collected in the 1990s, and, unlike the Norfolk inventories, material was extracted verbatim, and therefore provides more consistent and detailed evidence than we have for Norfolk. Forty-eight inventories in the Cornwall sample include references to 'moieties' or halves of livestock and crops, while a quarter of these also included moieties of farm equipment and household stuff.[21] For example, Richard Cooke of Madron, in 1617, besides stock, stuff and equipment which he fully owned, left 1 moiety of a 2 year old heifer, 13s 4d; 1 moiety of ½ year beasts and advantage, 26s 8d; 3 moieties of horses, 50s; moiety of mowhay and corn £4; 8 moieties of pans, 45s; 1 moiety of tableboards and 1 moiety of cupboards, 20s; 1 moiety of little tableboards and 1 moiety of forms 12d; and 1 moiety of furze ricks 6d.[22] This style of halving resources, livestock and crops conforms to Type E in our taxonomy. It is difficult to ascertain whether, or how far, this shared ownership equated to farming to halves. But the reference to 'advantage' alongside the ½ year beasts, may be a term for the uplift or profit on the beasts. Another inventory of Richard Davye, yeoman of Zennor, in 1647, records the bequest of '1 halfendeale' (Type A) of a tenement with chattels to his son, with a moiety of the growing corn.[23] The inventory of Maddren Trembath of Madron, in 1616, includes the reference to sheep 'upon half crease' (Type B).[24] Significantly, it was not listed in the body of the inventory, but appeared with the wool as a debt 'dew to the house'. In this case, there was a specific need to identify the location and status of the sheep; normally, there would only be the requirement to state the proportion of the animals owned by the deceased. In both these instances, evidence was slight, but sufficient to suggest that forms of sharefarming underpinned the moieties.

The 36 inventories without moieties of household goods and farm equipment, were heavily skewed towards livestock; only two included corn, one of which contained the 'halfendeale' agreement.[25] Cows and their followers formed the largest category by far, found in 26 inventories. The prevalence of dairy herds points to the existence of a system of sharemilking with the sharing of calves, akin to that described in the county survey of 1815.[26] Wills from other counties show that forms of cow hiring and cows at halves were common in pastoral areas. In south Shropshire, in 1552, John Mytton of Kinnerton had his kine and heifers set to parts and to hire, while cows at halves existed in Lancashire in the 1650s.[27] The practice enabled smallholders to obtain animals cheaply, while at the same time reducing the owners' need for pasture.

There are only two unambiguous examples of sharefarming in the Kent sample of inventories. In 1699 the appraisers of the inventory of Henry Austen, a yeoman from Wye, recorded 21 sheep and 20 lambs, but added, '21 sheep being in keeping with Lawerence Austen for half profit and one half of their 20 lambs'. Nicholas Adams, a yeoman from Lenham had 14 sheep 'to halves' in 1664.[28] There are other inventories where half the value of livestock or crops is given, and while it is possible that this reflects some form of sharefarming, as with some of the Cornwall inventories, in the absence of further information it is impossible to be certain. For example, John Ponnet of Wye had '4 acres of oats of which Edward Burges did to have half'. This could refer to some sort of partnership agreement, but could also indicate farming to halves.[29]

There are other scattered examples of farming to halves appearing in probate inventories. A handful of inventories identified by Thirsk from Yorkshire, Cumberland and Northumberland dating from the 1590s to the 1630s include references to animals at thirds and halves, and half shares of corn and a beehive. There are some examples however, which date from later in the century, for example, at Harwell, where William Bowldry of Streatley, in 1675, left 9 acres of wheat and 7½ acres of barley 'at haluese' and John Laurence who sowed to halves 10 acres in Long Wittenham field in 1602.[30]

How much significance should we attach to the evidence from probate inventories? The relatively few examples suggest that shared ownership and farming to halves was an occasional rather than a common practice, and not an overwhelming feature of village life.[31] However, the likelihood is that the picture is incomplete. The purpose of inventories was to value moveable goods to enable the executors of an estate to pay legacies and debts. In recording the value of items appraisers would only occasionally bother to explain their valuation, or put it in a wider context. Although no examples of farming to halves were noted in Suffolk inventories, a deposition from Sudbury in the late seventeenth century refers to farming to halves as 'the usuall way or manner of halving'.[32]

Despite the few inventory examples from Kent, Chalklin claims that 'sowing to halves was more common than payments in kind and special labour services'. Chalklin provides a definition of farming to halves in Kent: 'it was a type of tenancy in which the rent was partly in the form of labour and partly in kind. The landlord provided the ground free and half the seed, and he and his tenant shared the crop after harvest'. Thus, 'in 1604, John Austen of Northbourne, yeoman agreed with Simon Lott, of Sholden, husbandman, that Lott should sow to halves some land which Austen leased from Sir Thomas Peyton. Lott was to plough the land while Austen was to allow him a bushel an acre of seed; when Lott had harvested the corn they were to share it between them'.[33] This provides a classic example of farming to halves taking place in the context of subletting, which is where we would expect to find it, but where it is so seldom recorded. Venn, in one of his 'few recorded instances' of *métayage* in England, also cites a Kentish example.[34] At Knole, the 5th Earl of Dorset, accorded four farmers the right 'to plough anywhere in the Park, except in the plain set out by my Lord and the ground in front of the house, and to take their crops, and it is agreed that a ⅓ of each crop ... shall be taken and carried away by my Lord for his own use'.[35] As far as Venn is concerned the agreement amounted to nothing more than the temporary leasing of grassland for tillage, but, he accepted the basis of the payment was essentially that followed in modern Continental *métayage*.

A similar awareness of *métayage* can be found the *History of the Parish of South Ormsby cum Ketsby*. Massingberd, writing in the 1890s, quoted a lease granted by an ancestor in 1683 as 'an interesting example of the Métayer system of letting land existing in the 17th century in Lincolnshire'.[36] From our perspective, it is significant that he was familiar with the practice, able to identify it in the documents and regarded it of sufficient interest to set out in full. The agreement between Drayner Massingberd and the labourer, William Brookes, is unusual in that it combined arable, wool and lambs. For the arable, Massingberd provided the land and seed to be sown, while Brookes agreed to live on the farm, repair fences, plough and sow the seed, mow, reap, turn and cock the corn, make the hay, and look after the goods and chattels on the farm. In return for 'faithfully performing' these tasks, he was allowed an eighth part of the corn, to keep two kine in winter and summer on the farm, and one beast. He was also to have an eighth part of all the lambs and the eighth part of the wool of the sheep kept on the farm. Massingberd provided the oxen and horses and two more horses in seed time, and agreed to pay 'for the shoeing of the horses and the mending and repairing of the ploughs and geares and harrowes, and to pay or allow for all taxes and assessments to the constable, church and poore and the townes shepherds wages'. Massingberd clearly got the better of this bargain for the next year on taking in hand another farm it was agreed that Brookes should have the sixth instead of the eighth part, 'for his pains and charges', for the whole land under his charge. This lease most resembles the agreements Sir Roger Townshend of Raynham made with his shepherds in the 1630s, in that Brookes provided nothing but his labour, and had no share in the working capital of the farm. In his account book, Massingberd made a variety of similar agreements with small tenants and labourers, including several with Robert Toppin in the 1660s 'to plough my arable to halves'.[37]

In his work on rural credit, Holderness also refers to *métayage* which existed alongside the deferred payment of rents and loans for the purchase of livestock as a form of credit extended to tenants by landowners and the clergy.[38] 'The uncommon English version of *métayage* – halves and thirds – is best seen as a credit transaction to enlarge the farmers' capital or to improve his liquidity. None of these expedients laid a permanent incubus of peonage upon English rural society, nor vitiated the landlord-tenant relationship in the manner of Southern Europe, partly because they were so few, but chiefly because they always remained short-term casual adjustments'.[39] This assessment, in the light of the present research, requires some modification, yet it shows an understanding of the practice by a historian with a Lincolnshire background, confirming the observation made by Massingberd in 1892 that *métayage* was particularly associated with that county.[40] The biography of Thomas Jackson of Aston by Budworth in Cheshire, shows how these kind of credit arrangements, mentioned by Holderness, worked on the ground.[41] Jackson, trained in the law and land surveying, served as steward to Sir Peter Leicester of the Tabley Estate from 1649. Amongst the estate papers Jackson kept a record of his own concerns, which included a small farm he inherited from his mother. His notes 'touching my estate' in 1654, which he drew up when his lease of three lives was renewed, record precisely £9 5s rent for the farm, ten cows let for 10s a head, fields leased singly and re-let to the highest bidder for £2 6s 8d and money out on loans totalling £35 16s. In 1656, he also noted 'the moity of ye Pease in ye Loontfield to my part was 18 thrave within a sheaf'.[42] These notes are of the greatest significance; they show an educated man of comfortable, but limited means, subletting his farm, lending money to his neighbours, hiring out cows and sowing his arable lands to halves. It illustrates to a nicety the varied farming portfolio held by an enterprising yeoman, with an acute understanding of the possibilities of making money from the land; in his calculations forms of sharefarming played a prominent part.

Further examples indicate that at this social level, where medium sized landowners were able to exert control, or wanted to maximize their returns, farming to halves was a standard option. Amongst the Verney papers, an agreement from East Claydon in Buckinghamshire, shows John Duncombe, gentleman, leasing his arable lands in the common fields to halves in 1651.[43] The agreement is set out as a formal lease and the term was for six years. Thomas Hughes, the tenant agreed to 'plough, manure, sowe and dresse the said land', Duncombe provided the dung, while they shared the cost of seed. One week before harvest, they made a division of the land to be cropped 'by lot', the one half to the 'use and benefit' of Duncombe, the other half to Hughes. In addition, Duncombe provided grazing for his horse and a cow, and sheep to fold the lands, while Hughes agreed to cart wood and fuel and to pay towards the wages of a shepherd. At Elmswell in East Yorkshire, Henry Best made an agreement with John Whitread of Little Driffield in 1625, 'for 4 oxgangs of corn which was my half part of Joane Wise's farm'.[44] Across the Pennines in Lancashire, the Walmesley family, unable to find tenants during the crisis of 1622–4 sowed lands to 'half parts' at Dunkenhalgh.[45] In Cambridgeshire, in 1589, Richard Biddall of Willingham, who made a good living from 2½ acres, directed in his will that his brother should manage his holding for half the profits until his nephew came of age.[46] This was precisely the arrangement inherited by Robert Loder in 1610.[47]

The unifying theme linking these examples of farming to halves, sharecropping and cow hiring is the association with experimentation, development and improvement, particularly in the first half of the seventeenth century. Forms of sharefarming offered a flexible way of farming, sharing risk and accumulating capital for both landowner and tenant in a period of opportunity but also uncertainty. The price rise of the late sixteenth century, which saw corn prices rise five fold, had created a new and dynamic market in land and agricultural products.[48] If landowners were to benefit, they needed to commercialize the management of their estates, exploit their demesnes and bring waste and unimproved land into cultivation. At the same time, the price rise offered the enterprising and skilful yeoman every incentive to experiment and expand his business. For both parties, as Kerridge and Holderness described, farming to halves provided a low-cost way of financing new enterprises in a period when capital formation amongst farmers was still weakly developed. In 1652, Samuel Hartlib had pinpointed shortage of capital as an obstacle to improvement; he recommended that industrious gentlemen, 'to encourage some expert workmen into the place where they live, should let them land at a reasonable rate and if they be poor and honest to lend them a little stock'.[49] Farming to halves performed that kind of function.

European examples reflect a similar strategy and chronology as landowners and farmers, responding to the price rise and economic turbulence of the sixteenth and seventeenth centuries, commercialized the management of their estates. In France, *métayage*, little known in the medieval period, increased rapidly in the seventeenth century. In the west of France, for example, in the region known as Gâtine Poitevine, share contracts rose from less than 5 per cent in the 1580s to over 90 per cent a century later. By the late eighteenth century, they remained rare only in the north and east of France.[50] In 1600, the French agriculturist Olivier de Serres recommended such a course as being more profitable than charging a fixed rent, in that the landowner did not have to concede such a large reduction to the tenant for taking on the entire risk.[51] It also reduced the chances of default, as he could go along at harvest time and collect his due. At a time when tenants might be insolvent, lack the necessary capital or hesitate to shoulder the risk, letting land on shares became a popular solution. As in England, in this transitory phase from medieval subsistence to a fully fledged capitalist agriculture, sharecropping played a vital role easing tenants into new holdings, and sharing the cost of improvements. In Portugal, Santos explains how landowners resorted to sharecropping during the difficult years of the seventeenth century, having entered into fixed rent contracts in the expansive years of the sixteenth century.[52] The likelihood is that there was a similar trend in England, particularly in those areas, like Norfolk, where radical and risky improvements were being undertaken.

5
Seventeenth-Century Case Studies: Farming to Halves on Four Norfolk Estates

The purpose of the Norfolk case studies presented in this chapter is to describe the context in which sharefarming existed and developed during the course of the seventeenth century and into the eighteenth. Sets of agreements from four Norfolk estates show why landowners resorted to these methods, how they adapted them to particular circumstances, the problems they encountered, the success they achieved, and why they abandoned them in the early eighteenth century. The case studies, drawn primarily from estate records, inevitably present farming, or letting, to halves from the landowners' perspective. However, they still provide insights into forms of sharefarming at a lower social level, between tenant and sub-tenant, yeomen and husbandmen, and generally within rural communities. Norfolk landowners used farming to halves throughout this period, firstly, as a device for effecting expansion and improvement from the 1600s, and, secondly, as a way of dealing with vacant farms and falling rents from the late 1660s. The driving force behind the improvements of early 1600s, was the inflation of the late sixteenth century. Prices accelerated rapidly in the 1580s and continued rising until the 1640s, leaving landowners in a vulnerable position. To maintain their incomes, they needed to modernize their procedures, place their rents and tenures on a commercial footing, and exploit their demesnes. If they failed to do this, potential revenue would drain away into the pockets of tenants and endanger the survival of their estates. In the second half of the century, as prices fell, landowners were forced to revise their strategies once more. They needed to diversify from corn to more profitable stock farming, and intensify their farming regimes. On all four estates, the owners collaborated with their tenants, leased dairies, sowed arable to halves, subsidized corn production, and assisted with the provision of working capital. Detailed sharefarming agreements from

Raynham and Felbrigg, and variants from Blickling and Hunstanton, allow us to examine how the practice worked, how agreements differed, and why they succeeded on some estates and in certain situations, but not in others. Assessments will be made of its operation in different parts of Norfolk, the impact of individual landowners, bailiffs and stewards, and its significance on these Norfolk estates.

For each of the four Norfolk estates the early years of the seventeenth century represented something of a fresh start.[1] Raynham and Hunstanton were well established estates undergoing a radical programme of improvement, while Blickling and Felbrigg were passing into new ownership. In each case the process of acquisition, expansion and improvement generated huge quantities of estate records. Activity stalled in the 1650s and 1660s, only to revive in the early 1670s, as landowners adopted the new initiatives associated with maintaining rent levels. In the late 1690s, the owners withdrew from day to day involvement, and simplified their estate management. From that time, estate records become less informative and simply told the landowner what he needed to know, reminiscent of the changes when landlords leased their demesnes after the Black Death. The evidence of farming to halves varies in quality and quantity, depending on the particular needs of landowners and their management style. For example, no references to any form of sharefarming were found at Blickling in the first half of the century. Surviving material consists primarily of surveys, valuations, leases and formal accounts, as Sir Henry Hobart set about reorganizing acquisitions, reconstructing farms, raising rents and investing in buildings and infrastructure.[2] His idea of improvement meant creating large, well capitalized farms which he could let to 'honest and able' tenants, often landowners in their own right and sometimes less fortunate relations. While he required them to look after his investment, repair the buildings and care for the newly planted orchards, he left the actual farming operation to their own discretion. Hence there is little information on farming activity on Sir Henry's new Norfolk estate, although this does not mean that sharefarming did not occur under the aegis of these large tenants. His successors, led by a most astute estate steward, implemented various schemes to keep these tenants afloat when prices turned down in the 1670s. At the other end of the scale, Thomas Windham, who inherited a much truncated estate in 1616, developed an intricate livestock fattening enterprise with local butchers and graziers, which can be compared to the activity in the Welsh Borders. Unfortunately, his record keeping was poor, so the picture is incomplete. However, his son, as we have seen, left the most detailed evidence of letting to halves in his estate book in the 1670s and 1680s. At Raynham, glimpses of the practice appear in the 1620s and 1630s, linked to new initiatives and better documentation; they fall away during Sir Horatio Townshend's early years, and then increase markedly from the 1670s as he pays the closest attention to the management of his estate. Of the four families, the Le Stranges kept the most comprehensive records of their farming and estate management in the first half of the seventeenth century, and provide the fullest references of farming to halves at that time. However, the quality of the material declines in the second half of the century, partly the result of two fires at Hunstanton which destroyed much evidence from that period. It is significant that where and when the landowners kept the most detailed records, evidence of farming to halves comes to the surface. For this reason, the case studies will start with the Hunstanton estate, which will give a detailed description of the practice, and may well indicate what was happening elsewhere.

Sharing risk and effecting improvement, 1600–1650

The Hunstanton estate lies in the north west corner of Norfolk, abutting the Wash to the west and the north Norfolk coast to the north. The fringe of coastal marshes running from Heacham through Hunstanton to Holme, and the sandy brecks rising to the east in the parishes of Ringstead and Sedgeford, offered much scope for improvement. Hunstanton Hall, with its park and demesne lands lies at the centre, nestled in a fertile river valley. In all these situations the Le Stranges farmed to halves. Sir Hamon had inherited the estate when he came of age in 1604. It had been in the hands of trustees since his father's death in 1592, and was by all accounts severely rundown, as his wife Alice noted; 'my husband was left in debt by his fathers Executors, and was left neyther stuffe nor stock, and his chiefe house halfe built, and all his farme houses in such decay. So he hath built most of them out of the ground.'[3] At the outset, with several annuities to pay, their rental income amounted to little more than £700 a year which forced them to borrow to cover expenditure.[4] The situation was precarious, and without supreme effort the estate might have succumbed like so many others at this time. Sir Hamon set about his task with vigour, commissioning surveys of the entire estate, drawing up field books, rentals and farmalls,[5] equipping himself with a full picture of the landholding structure and topography of the estate; he also kept several memoranda books.[6] From 1609, he gradually handed over the household and estate accounts to Alice, who proved to be a skilled accountant and manager. She reformed their administrative procedures, modernized the accounting systems, and provided Hamon, and later their eldest son, Nicholas with the management support they needed to carry out improvements. Nicholas inherited his parents' diligent habits. In his notebooks he kept detailed evidence of farming to halves on the newly drained marshes.

The first example of farming to halves actually predates these initiatives, and indicates that the practice was already well established on the estate. In the bailiff's account for Ringstead, written in English, is the note, 'The profitte of the tythes and halfes of the Corne growing in the field of Barrett Ringstead in the yere 1605', which lists the seed delivered, corn threshed and grain sold.[7] The background to this arrangement is similar to the agreements made at Raynham in the 1470s and 1480s, which involved the cultivation of corn on the breckland. Barrett Ringstead is a deserted village which lies to the south of Hunstanton and abuts the park. Sir Hamon leased the foldcourse to several farmers, in fact four sets of partners, who paid £25 each (Type E in our taxonomy).[8] He retained 'all the grain and corn growing there …' and sowed the corn to halves, as the Townshends did in the fifteenth century, corresponding to our Type A.[9] In his book of receipts and disbursements, he noted in 1607, 'Rec of Roger Browne £34 for last paym of £64 wch he was to pay for my part of the halfes in Barrett Ringstead; out of wch I allowed him backe 20s for the halfe of 5 acres of wheate in the use of Mr Crotch'. Payments for threshing and golfing (stacking) 'my halfes' continue from 1607 until the 1620s; for example in 1607, 'payde to Thomas Savadge for 20 dayes worke with me in harvest for golfing of my halfes at 8d the day', and in 1622 'to Andrew for thrashing of 5^{xx}9 comb of barley … for his halfe.'

Why did Sir Hamon enter into this sort of agreement? Why did he not let the land or employ direct labour to cultivate it? The Le Stranges, like the Townshends and the Pastons, had long traded in grain, but they did not cultivate it directly, nor did they possess the equipment or the management structure to deal with it. At Sedgeford, as we have seen, they relied on a mixture of corn rent (Type H) and money rent, thus easing the risk for the tenants on the sandy brecks, relieving them of the business of cultivation and allowing them to benefit from rising prices.[10] With the principal tenants already engaged in the collection and disposal of large quantities of tithe corn, a proportion of farm rent paid in corn made sense. The management of the Barrett Ringstead foldcourse was similar. Its marginal viability meant that the foldcourse had to be leased to a syndicate of farmers. By sowing the arable to halves, Sir Hamon ensured sufficient seed for cultivation and benefited from any upswing in the market. In this way Sir Hamon retained flexibility and control over cultivation.

Farming to halves was not only used on the brecks at this time. In the Hunstanton Farmal, Sir Hamon, in a notebook of farm rents, recorded two agreements in 1613.[11] The first was with his butler, Thomas Ketwood, who 'sowed of barley to halves in the Upper Close 10 [combs] in the Lower Close 5 combs, for Gotterson and himself, so that the whole being 15 [combs] it must be 7 [combs] 2 [bushels] for the halfe in discharge whereof I undertake to promise to satisfie Tho. Gittings of 5 [combs] wch Ketwood oweth him and 2 [bushels] I allow him to be rec of William Cobb wch he had of me than used concerning my halfe seed and for the other 2 [combs] I allowed him in the former note on this page'. Secondly, 'Will Cobb sowed of gray pease and fetches to halfes with me in Poole Close …'. In 1614, he paid for 'barley to sow the halfs with William Cobbes at the 10s 6d the comb'.[12] Both these agreements concern enclosures near Hunstanton Hall and appear to be short-term contracts to plough up and improve old pasture, in the manner suggested by Kerridge.[13]

Perhaps the most significant examples of farming to halves were those concerned with the reclamation of marshland. This project started in the early 1630s with the marriage of Sir Hamon's eldest son Nicholas, possibly as a venture for the young man eager to expand his inheritance. Following the example of his parents, he kept meticulous notebooks, of which four survive, detailing his activities from 1631 to 1653.[14] From the outset, the layout of the notebooks indicates a carefully planned enterprise, starting with embanking and the draining of marshes, progressing to cultivation and to the construction of farmhouses, equipped with specialist buildings, gardens and orchards. The entire operation was costed and analysed using mathematical tables and almanacs. Payments for labour and materials do not appear in Alice's accounts, so we can assume that Nicholas kept his own day to day accounts, which have not survived. The scale and complexity of the work required him to draw up a table of 'Rates and Wages in Draining', as he employed an army of labourers, craftsmen, smallholders and contractors to perform the huge range of tasks involved. He entered into various agreements with specialist workers, utilized their equipment and paid incentives for good work and extra effort, for example, 'A largesse to my wettworkemen, upon bargaines, or working at unreasonable houres in cases of necessity, for taking up dames, or the like'.[15]

It was when Nicholas came to cultivating the Bogge in 1637, that he resorted to 'ploughing & to halfes'. The Bogge (23 acres) and Meeles (16 acres), which lay near Hunstanton Marsh, were leased by William Murton and John Woodrow, but behind this tenancy lay a further agreement whereby Nicholas shared the cost of ploughing, sowing and harrowing the newly improved land. The account lists payments 'for my part'. For example, 'to the townsmen's men and sonnes that helpt me to plough and harrow my part … for my part of seed wheat … seed rye … for keeping away vermin … for watergrupping the ground.' This arrangement continued until 1643, when Nicholas rented the Bogge from his father and used direct labour. Ploughing to halves was not confined to the Bogge. For 'the Whinne Pasture, New Close, which ground I halved with Rich. Gyles', he also shared the cost of levelling and carting, 'the tumbrels, 4 labourers and 3 horses about that place five pounds five shillings in which Ri: Gyles bare half, so my part comes to £2 12 6'. Nicholas made these agreements with existing tenants as a way of sharing the risk and cost of improvement. On the marsh itself, which he leased from his father, he employed direct labour, as he did with Bogge when he took over the tenancy in 1643. Without Nicholas's notebook, we would have no idea that this arrangement lay behind the formal tenancies.

In 1637, Nicholas also shared the cost with his father of improving Holme New Marsh Close, 'in this peece of ground my father bare halfe the charge of leveling, fencing, ploughing &c'.[16] Sir Hamon's shares of the payments were entered in the household accounts from 1638 to 1643.[17] For example, 'Mar 1640, Holme Close, Reaping the Half Charge', included 'for weeding the close, the halfe, for twice drawing and cleansing the river.' When the 'halfe charge' was not noted in the margin, the clue is that the cost is halved, 'for mowing 9 acres with Barley and fitting it to the cart at 2s 6d per acre, 11s 3d'. A later notebook records another agreement between father and son for the Holme Marsh from 1643 to 1645, 'This ground was set out by Mr. Hasdoncke the Dutchman as part of my father's proportion due to him upon the embanking of Holme Marshes, and used by my Fa. And Me, in a PARTIBLE way of charge and profit'.[18] The employment of such a man, with the latest Dutch equipment, shows how innovative the Le Stranges were. In this case, Nicholas provided half the seed rye, wheat and oats, and paid 'my part' for ploughing, harrowing, and sowing. This arrangement between father and son may have been influenced by the use of the marsh for 'gunning', a shared passion; the work included the construction of 'breast works or blinds … to gaine shoots at fowles'. In 1633, Nicholas had constructed three osier islands in the Marsh Poole at Hunstanton, for 'fish, fowling and brooke-hawking'. At Holme Marsh, in 1646, for the osier grounds and islands, Nicholas agreed with Makemede and Write 'to beare halfe the charge of cutting and bunching' the osiers, an arrangement which continued until 1653. This agreement may reflect the risk entailed with these marsh crops; in the account for cutting reeds, 'the charge of sheering was drowned and swallowed up in the profits made of it.'

The final notebook conveniently provides a summary of the work undertaken by Nicholas and confirms the interlocking nature of the programme surveyed by John Fisher and improved 'by me'.[19] This summary can be cross-referenced back to the earlier notebooks, Alice's accounts and the maps of Hunstanton, Sedgeford and Heacham. The tables indicate that much of the detail on Heacham has been lost in the missing notebooks, and that the acreage drained far exceeded that at Hunstanton.[20] Alice's accounts provide glimpses of activity, which include sowing to halves the arable cultivation and looking after wintered cattle. For example in June 1643, 'Layd out as appeareth by Heacham Marsh reckoninge the halfe of the charge for Bearne farme stacks, removing, thrashing, looking at the cattle wintered there and the halfe of 2 earths ploughing 21 acres 2r and 22c of seed beanes at 6s 8d ... £20 14s 7d.' And in July, 1645 'Layd out as appeareth by sonne Strange his reckoning for reaping, thrashing and the Bearne farme for the Beanes & winter corne in Heacham Marsh 1644, the halfe .... £4 19s 8½'.[21]

Taken together, these notebooks provide evidence of a sophisticated scheme of land improvement. It is in this context that we find the systematic and documented use of farming to halves as a tool for improving the landscape, extending cultivation and developing amenity areas on the estate. This concern with the aesthetic as well as the functional remained the hallmark of all the improvement undertaken by the Le Stranges from 1606 to 1654. It is a surprisingly modern concept, anticipating the landscape movement of the eighteenth century, and the management initiatives of today. From our perspective, farming to halves played a significant role at Hunstanton, supporting tenants during the initial risky phases of marsh drainage. It tackled, successfully, the issue of finding and establishing tenants, helping them to accumulate working capital, giving them the means to overcome a few lean years, and the ability to stock and farm their holdings. Their success compares with the repeated failure of Sir John Hobart's tenants on the fenland estate at Stow Bardolph in the 1630s, and the short-lived cattle fattening scheme adopted by them in the 1640s.

At Raynham, in the early seventeenth century, the enthusiasm for land improvement was not so marked. Nor did the Townshends, at the outset, display the unity of purpose that was such a feature at Hunstanton. The estate had been through a difficult time following the death of Sir Roger Townshend in 1590. His son, John had wasted his fortune and left the estate in a parlous state; but for the efforts of his mother, buying in portions as they came up for sale, it might not have survived.[22] When he died in a duel in 1603, she took responsibility for what was left of his estate, and his eldest son, another Sir Roger. Remarried to Lord Berkeley in 1595, Lady Jane had secured a life interest in the major portion of the estate and continued to enlarge and improve it until her death in 1617.[23] Largely due to the reforms she initiated the Townshends possess enviable estate records; the bailiff accounts survive in an almost unbroken sequence from 1604 to the eighteenth century. In addition, the Sheep Reeve kept a full record of the twelve flocks kept by the family, and it is here, in the 1620s and 1630s, that we found references to farming to halves. The practice also appears fleetingly elsewhere around 'the Raynhams' and 'the Rudhams', where the Townshends concentrated their management energies. But beyond the core estate we found little evidence of the practice; this is probably explained by the management and the type of documents kept on the offlying properties, which concealed the farming activity that actually took place.

The six bailiwicks making up the estate included the lands at Raynham cum Membris: East, West and South Raynham, Pattesley, Helhoughton and Shereford, the newly acquired estate at East and West Rudham to the west, properties at Stanhoe, Barmer and Barwick and South Creake to the north, and the coastal estate at Stiffkey, Langham and Morston, which Sir Roger Townshend inherited from his mother, Anne, the daughter of Nathaniel Bacon, in 1630.[24] To the east of Raynham, lay the small estate at Stibbard and Ryburgh, where farming to halves occurred in the fifteenth century.[25] Lady Jane's jointure lands included the Raynhams, the Rudhams and South Creake, while John's widow, Anne lived with her children at Stiffkey.[26] In total the estate was worth just over £2,000 a year.

Although rarely resident at Raynham, Lady Jane's records show her commitment to estate improvement. For her benefit, the bailiff accounts were written in English from 1604 before reverting to Latin in 1617, and contain much explanatory detail.[27] Her strategy centred on the management of the sheep flocks and the enlargement of existing holdings, notably at the Rudhams.[28] In 1579 and 1588, her husband had purchased two manors in East Rudham, including Coxford Priory, from the Earl of Arundel.[29] Lady Jane acquired two more in West Rudham, the Great Grounds with its foldcourse in 1609, and in 1616 St. Faith's Manor, thereby doubling the rental from £220 to £450 a year.[30] Four sheep flocks, grazing the extensive foldcourses, underpinned the fertility of these light sandy soils. The value of the property lay in its proximity to Raynham and its potential for improvement, the risk lay in the need to maintain high stocking rates, attract well capitalized tenants, and the continuance of buoyant corn prices to finance the improvement. Without these ingredients, the whole enterprise could unravel, as it did in the mid-1660s.

The earliest accounts of the Rudhams indicate the fragility of the farming regime with the brecklands 'let by the year to divers psons for 2/- the acre', and a proportion of the rent paid in barley.[31] By 1617, a farmall listed over 78 tenants, in the four manors, paying between 4s and 5s the acre.[32] By 1619, 'a note for the brecks in the Rudhams' shows the brecks in Great Ground and St. Faiths let 'with tathe' (manure), for 6s 4d the acre, and sown with rye and oats.[33] This improvement was associated with the policy of taking the foldcourses in hand and restocking the flocks. The flocks included three in the Raynham area: Robinhoods, Kipton and Shereford; four in the Rudhams: Coxford, South Grounds, Mannihouse (St. Faiths) and Great Grounds; and one in South Creake. The Sheep Reeve account of 1605 recorded 6,200 sheep.[34] Numbers remained more or less at this level until 1618, when they slumped to 2,546, leaving Sir Roger with the task of rebuilding the flocks.[35] By 1622, numbers had risen to 9,101.[36] In 1623, with the additional flocks from Barwick, Barmer, Stanhoe and Stibbard, they reached the staggering total of 12,110. It is in this context we found the first reference to farming to halves with shepherds being paid a portion of the lambs and wool, 'shepherdes partes', alongside the payments of the tithe.

'Shepherdes partes' first appear in the account drawn up by Thomas Barker, Sheep Reeve for Sir Roger Townshend, for 'the lambes and woolle rennewinge upon his several foldcourses ... 1622'.[37] This account supplemented the General Account of the Sheep Reeve which recorded the sales of wool, lambs, skins and crones, and expenditure on wages, hurdles, tar, pitch and other necessaries for the sheep. After recording the charge of the sheep in his care, Barker noted the tithe delivered to the various parish incumbents, and then the shepherds' parts:

Shepdes parte delivered to the shepde of Kipton for his xxiiii pte of 6C 4^{xx} [680] lambes renewing there this yere besides the tithe, xxxi and a half a lambe. To the shepde of Robinhoods for his seventh parte of 3^{xx} vii [67] lambes renewing there this yere besides the tithe, ix. To the shepde of East Rudham south grounde for his nynth parte of 7C liii [753] lambes rennewing there this yere 4^{xx} xix [99]. To the shepde of Mannihouse for his nynth parte of 7C 3^{xx} xviii [778] renewing there this yere 4^{xx} xix [99], besides ii lambes allowed to the lord because the lord had in the same grounde 34 ewes above 9C. To the shepde of Shereford for his tenth parte of 2C 5[xx] 10 [310] lambes renewing there this yere beside the tithe xxxv. in all 2C xxxv 10 and a half

Marking lambs delivered to every of the Shepdes of the aforesaid viii several groundes ii lambes for marking as formerly hath been allowed. in all xvi

And so remayne in Sheep Reeves hands this yere to be sold and to be layd out on aforesaid severall groundes 3M 1C xxiii and a half [3123½]

The reference to marking lambs, 'as formerly hath been allowed', indicates that the shepherds parts were a new initiative concerned with the increase of lambs. By offering a share of the output, shepherds were encouraged to look after the newly expanded flocks and to maximize lambing rates. Townshend also allowed the shepherds' parts of wool:

Shepdes partes of woolle delivered to some of the shepdes for their severall partes that is to say to the shepde of Kipton for his xxiiii parte, iiii stones xi and a half, to the shepde of Robinhoodes for his seventh parte xii stones iiii poundes, to the shepde of Lodge Flock for his eighte parte xvi stones besides some woolle allowed because the lord had more then 8C sheepe. To the Shepde of West Rudham great grounde for his tenth parte xiiii stones besides some woolle allowed because the lord had more then 10C sheepe. To the shepde of East Rudham south grounde for his ninth parte x stones. To the Shepde of Mannihouse for his ninth parte besides some woolle allowed because the lord had more then 9C sheepe. To the Shepde of Shereford for this seventh parte vii stones in all amounting to 3xx xv stones 1 pound and half a pounde.

Wooll remayninge   Soe remayne of the wooll of this yere delivered
                   unto the wooll house 6C xxxvii stones v pounds

In addition to the parts of lambs and wool, the shepherds received wages, although references to 'shepherds parts' continue through the late 1620s and 1630s, sometimes referring to lambs or wool and sometimes both.[38]

William Stanhowe included references for both lambs and wool in his Sheep Reeve's Accounts for the 1630s, following the abridged format evident in 1627 and 1630.[39] The entries end in 1639 when the trustees let the flocks. There are no references to 'shepdes partes' in the surviving accounts for the Stiffkey and Morston flocks in 1633, 1634 and 1635.[40] This could mean that the practice did not occur here, or simply that it was omitted. Alternatively, the practice might be associated with particular bailiffs. Stanhowe had been bailiff for East and West Rudham since 1632, and we find in 1634, in a receipt attached to the back of the account, 'paid to Wm Grove his parte of three ockers of otes he did sow to halfe with my master at West Rudham breck and he was to halfe the corne by his agreement and the otes newe laid into my msters barne and he hold the otes so due to me witness Christopher Sella'.[41] There is no reference to this agreement in the body of the account, and at the same time, no indication that the arrangement was in any way unusual. It seems that occasionally it occurs to the bailiff to enter this type of information, but it was not common practice to do so. These examples suggest that references to farming to halves were explicit when applied to new initiatives like the expansion of the sheep flocks at Raynham, and the draining of the marshes at Hunstanton, but remain more or less concealed otherwise. The nature of the documentation and the style of management remains vital in determining whether farming to halves comes to the surface. In the account book rendered by Edward Symonds to the trustees from 1636–1643, which needed to be very detailed, we find another example, 'rec'd of Cooper, 9th August 1637 for the moytie of 2 acres of Rye wch he did sowe to halves at Kipton the sume of £2'.[42]

Further accounts confirm that the early 1620s was a period of renewed activity and improvement. From 1621, the bailiff of husbandry at Raynham was instructed to keep a 'distinctive book of payments', which show how the labour was organized and the work accomplished.[43] He had to present the book to the steward for signing, who made a monthly abstract of it, reminding us that a layer of accounts existed below the level of those presented to the landowner. The payments, including wages for all sorts of task work and day work, show that direct labour was used extensively to cultivate land, sow crops, hedge, ditch and carry out repairs. There is no suggestion of farming to halves. This evidence, with that of Hunstanton, indicates that farming to halves was not used as an alternative to direct labour, but as an informal arrangement with existing tenants, often for a specific purpose, or to provide incentives to skilled and valued employees. It is significant that the references come from the sheep flocks in Raynham and the Rudhams where activity was concentrated and the records most descriptive. Further afield, the picture is more obscure.

When Sir Roger Townshend succeeded in 1617 the accounts for the offlying properties at Stibbard, Ryburgh, South Creake, Barmer, Stanhoe and Barwick returned to the main estate. From 1631 when William Clarke became bailiff, these accounts included only the barest detail of holdings, which hardly alter from decade to decade. Some of the early accounts indicate that this format concealed subletting, as in 1632, 'Monies due from Mrs Bunting for the farme of certain lands in Stanhoe in her farmers use for 1 year ended.' At South Creake, Mrs Sinckler 'demised to divers farmers at 5s the acre'.[44] In 1649, 'Sir John Tracey Kt for the Lo's foldcourse of Beaufois (South Creake) together with certain lands there in the use of the Thomas Linge and Robert Ling by the assignment of Sir John Tracey for one whole year'.[45] This example suggests large tenants acting as middlemen, making their own arrangements with sub-tenants and obscuring the reality of how the land was actually occupied and farmed.[46] A comparison with the accounts of East and West Rudham, where detail was recorded, shows that holdings were repeatedly divided, shared and reassembled. The likelihood is that William Clarke's accounts conceal this sort of activity between large tenants and their sub-tenants.

With the death of Sir Roger in 1637, large tenants became increasingly important as the trustees for Sir Horatio relied upon them. Like William Clarke, they often leased the foldcourses and assumed the role of bailiff. Peter Stringer, the principal tenant at Raynham from 1640, was appointed bailiff the following year and served until his death in 1665. In his account for 1649, he recorded the neats pasture divided between several persons, which is the first glimpse we have of cow grasses hired by a layer of cowkeepers.[47] In 1656 Stringer leased several foldcourses including Robinhoods at Raynham, Kipton at Helhoughton, the South Ground at East Rudham and kept accounts for 'the Stiffkey business'.[48] At the Rudhams, William Symonds, lessee of the Coxford and Mannihouse foldcourses, served as bailiff from 1641 to 1644.[49] John Robotham, lessee of Mannihouse foldcourse, arrived in 1649 and served until 1670.[50] At certain times, particularly when changes were being implemented, glimpses of concealed activity emerge. For example, the shepherds' accounts of 1647 refer to the 'Tathe: none, because it goeth into the Brecks to the halfe', in 1650 for Kipton foldcourse, the 'Tathe: it goeth wth the Brecks in halfes with Mr Stringer', and in 1651, 'With the quantitie of oats growing upon the Lords halfe pte of the brecks at Kipton and wth the Rye and Wheat growing thereupon the Lo's halfe pte of the breck'.[51] In 1652, William Clark, then Sheep Reeve, charged himself with the 'farmes of the brecks' ready sown with rye and oats for 10s the acre', indicating the ongoing need to intervene in the management of the brecks and foldcourses in West Rudham.[52] In the same year, Sir Horatio took Kipton and the West Rudham flocks in hand, spending £934 on 'cattle and sheep to stock the foldcourses'. By 1656, the flocks had been re-let to men of substance, like Peter Stringer.[53]

The situation at Raynham remained relatively stable until 1665 when Peter Stringer died and corn prices started a steep descent. In that year, the tenant of Coxford Priory failed, Stringer's foldcourse in East Rudham was divided between two tenants and the brecks of the Great Grounds in West Rudham were divided into three and let to tenants, ready sown.[54] At Great Grounds for example, 'the fearme of Edward Stott for 6 ac in the 18 ac Close £2; and for 9 ac & the 3rd parte of 2 ac of Gr. Ground Breck sown with wintercorne att 10s an acre £4 16s 8d; and for 7 ac & the 3rd parte of 2 ac of sumerley at 10s an ac £3 16s 8d, and for 10 ac sowne with Otes at 10s an acre £5, and for the 3rd parte of 5ac, formerly let with the Great Ground att 5s an acre £16 1s 8d'. At Coxford, where Townshend shared the cost of cultivation, 'The farm of Mr Parkins, respited the last year being now sett of wth him for seed & tilth... at Coxford £55... Jo Robotham for carrying out, filling and spreading of our halfe of 17 score and 12 lodes of muck at Coxford at 8d a lode the sum of £6 4s 8d'. At Raynham, William Clark supervised the reorganization of holdings following Stringer's death, which included 'the farme of Samuel Jervise for messuage & divers lands & meadows formerly letten att £40 by the year, but this last yeare butt att £19 by third yeare, in regard he doe sowe the Lords brecks and his own infield lands to the halfes with the Lords and this yeare the Lorde lands wth his own wch he formerly sowe to halfes nowe layed to his farme, he is to be charged att the some of £53 8s 0d'.[55] With Stringer in charge, the likelihood is that this arrangement would not have appeared, but Clark, less familiar with the bailiwick, needed to make a note of it. As these examples show, landowners like the Townshends remained heavily dependent on well capitalized tenants. When they died or failed, they were forced to resort to direct farming, dividing holdings and sowing to halves with sub-tenants, smallholders and 'petty farmers'.[56] As the economic climate worsened, the need for alternative strategies deepened and the evidence of sharefarming becomes more explicit.

Thomas Windham's situation at Felbrigg resembled more closely that of the Le Stranges of Hunstanton, in that from 1616 he was the resident landowner and took a close interest in the day to day management of the estate, negotiating with tenants and graziers, drawing up agreements and keeping his own accounts. Windham's primary objective was the development of livestock farming and the improvement of pasture. At Felbrigg this was achieved through the micro-management of the park and demesne lands, and the implementation of 'up and down' husbandry, as a matter of policy across the estate. Surviving leases confirm his commitment to the system, with restrictions on the ploughing up of light land and the insertion of grass leys to support increased stocking rates, and the regular cultivation of permanent pasture.[57] We found no evidence of him ploughing to halves, in the manner indicated by Kerridge, or sharing stock, as in the Welsh Borders, but he invested heavily in bullocks, providing the working capital and sharing the risks with local butchers.[58] Although his account-keeping was generally poor, he left an account book from 1615–1647, which gives details of the grazing agreements for his livestock enterprise.

The livestock enterprise was centred on the demesne at Felbrigg which formed a continuous block of 700 acres, mostly enclosed pasture and arable, ripe for exploitation and improvement. The accounts show that bullock fattening was intricate to manage and expensive to finance; it required a substantial outlay for the purchase of lean stock, and a long period before any return was forthcoming. Furthermore, trust had to be placed in dealers and intermediaries. At first, with his bailiff William Harmer, and Thomas Cutlacke, the ex-tenant of Felbrigg Hall, he bought and sold stock without the aid of a middleman, but the process proved risky and difficult. They needed to find a reliable outlet for the fattened stock at the appropriate time in order to close pasture for haymaking in May and June, and to receive new stock in the autumn. To rectify this problem, Thomas switched to a system of grazing agreements with local butchers, some of whom already leased meadows on the estate.[59] The arrangement allowed them to select and buy the stock, invariably Scottish runts from St. Faith's fair, or Yorkshire steers from Hempton. To ease their cash flow, the butchers sold their lean stock to Windham and then bought them back at an agreed rate, which included the cost of grazing. Set against the butchers' account was the value of butchered meat supplied to the Felbrigg household, and the purchase money for the next batch of lean stock. The bullocks were purchased in the autumn, grazed for a year and then withdrawn in lots from Michaelmas to Christmas; a supplement was charged for grazing until the following spring as it encroached on haymaking arrangements. When circumstances allowed, more stock was bought at the Whitsun Fairs.

The system worked well as it relieved Windham of the burden of stock dealing which was a specialist activity, and allowed him to concentrate on the grassland management for which Felbrigg was noted. The money basis of the contract stemmed from the cash received from sales at local markets and the need to purchase new stock. The enterprise, benefiting from a well developed marketing network, did not face the uncertainties of more remote and unstable areas, like the Welsh Borders, so there was less need to share the risks. Although Windham supplied the working capital for the stock, the ultimate responsibility rested with the butcher who selected the animals, took them to market and made the profit, while Windham received an agreed rate for his grass and winter fodder. This was not half crease, (Type B) as they did not share the profit, but his return averaged between 20s and 30s a head, with a stocking rate of a beast per acre, compared to rents of 10s to 16s an acre for arable land.

The arrangements for the sheep are more difficult to follow. Having fattened a few hundred sheep for the first few years, Windham sold the flock in 1624. We have no information on the sheep until Edmund Britiffe of Hunworth leased the foldcourse and warren in 1641. It is possible that Britiffe, who assisted in the selection and management of stock and corn, had enjoyed some informal agreement with Windham concerning the foldcourse since the 1620s. The appearance of Britiffe as a tenant and intermediary is significant. It is likely that he was responsible for the grazing agreements, while his son, another Edmund, drew up and monitored the detailed letting to halves agreements at Felbrigg in the 1670s. Sir John Hobart of Blickling also used Edmund Britiffe to advise on the management of his Fenland property, which had proved so difficult.[60] As at Raynham and elsewhere, these advisers were invariably men of yeoman stock, with the technical expertise to implement new initiatives and develop livestock enterprises. As in other parts of the country, farming to halves was part of their repertoire.[61]

The first part of this chapter, concentrating on the early seventeenth century, shows estate management and farming in a state of transition, as forms of tenure, dominated by custom, were converted to new commercial contracts, leases and agreements.[62] In their efforts to modernize their estates, landowners sought large well capitalized tenants to hire their improved farms for fixed money rents. But at this transitory stage, such optimal solutions were not always feasible, or entirely successful. In the first few years, tenants often required support in terms of capital and stock, while landowners sometimes needed flexible alternatives. In this context, farming to halves in its different forms, sometimes provided an answer.

Responding to adversity, 1650–1700

In the second half of the seventeenth century, developments in farming and estate management took place against the backdrop of the recent civil wars, continuing political uncertainty and a sustained depression in agricultural prices. Landowners, large and small, faced real hardship with oppressive taxation, falling corn prices, declining rents and failing tenants. Taxation did not diminish during the 1660s and 1670s, and reached unprecedented levels by the 1690s, accounting for, in some areas, a fifth of rental income. In Norfolk, as elsewhere, grain prices entered into a period of long-term decline from the early 1660s.[63] In 1664, wheat prices at Felbrigg and Blickling dropped from a peak of 36s a comb in 1647 to 16s. In the years that followed they showed no sign of recovery, falling to 9s in 1683. A brief rally to 14s was followed by a further collapse to 9s between 1689–91. Oat prices experienced a similar fate, falling from 12s a comb in 1661 to 4s 6d in 1664, dipping to 3s 6d in the mid-1680s. Barley, sustained by the demand for malt, maintained its prices more successfully. Prices ranged between 8s 4d and 12s in 1661, edged lower in the late 1670s, but did not fall below 6s until 1683. All corn prices staged a recovery in the 1690s, before retreating to lower levels in the 1700s. Prices for livestock and animal products remained relatively buoyant throughout the period, offering landowners a viable alternative. However, as we have seen, the high capital investment involved in setting up such enterprises made bullock fattening and dairy farming a difficult option for tenants even in the 1620s and 1630s; few could afford to make the change, or take the risk, without assistance from the landowner. It was in this context, of low grain prices and the need to diversify into livestock farming, that landowners resorted to sowing to halves and leasing dairy herds to tenants in the 1670s and 1680s.

The most explicit documentation on the need for change and improvement, during this period, comes from William Windham of Felbrigg. Soon after his marriage in 1669, he announced his intention to create an estate book that would contain all details of rents, leases, monies owed and paid to him, and what he had allowed in taxes, repairs and other incidents. The Green Book, as it was known, was designed to inform and educate, and includes helpful comments reminiscent of the records produced by the Le Stranges in the first half of the century.[64] Like Sir Hamon Le Strange, Windham also lost his father at an early age, and received guidance from notable men including his father-in-law, the merchant and agricultural improver, Sir Joseph Ashe of Downton, Wiltshire.[65] Sir Thomas Browne, the philosopher and physician, who had lived in Norwich since the 1630s, also advised and befriended the family.[66] Browne, a leading member of the Hartlib circle, influenced the tree planting schemes at Felbrigg, while Windham wrote the Green Book in the style of Sir Richard Weston's *Legacie to his Son*, published by Samuel Hartlib in 1652.[67]

The estate inherited by William Windham in 1669 had suffered neglect at the hands of his half brother, John, who died in 1665, and deteriorated further when the family lawyer, who had little understanding of agriculture or how to deal with tenants, administered the estate during William's minority. Letters between the young man and his uncle, Judge Hugh Wyndham, pinpoint the nature of the problems.[68] Arrears had been allowed to mount up and tenants received little guidance or help, leaving them with huge debts. This often necessitated forfeiture of the tenancy, as at Dilham Hall in 1658 and 1662, which was broken up and let to halves, the first instance found on the Felbrigg estate.[69] Judge Wyndham resolved the situation by appointing a local man able to negotiate with tenants, assess repairs, formulate and enforce agreements, and keep regular accounts. William Windham determined to take on the task himself, with a team of practical men, enlisting professionals when required, as his father had done. He differed in his emphasis on record keeping and accounting, no doubt influenced by his commercially minded father-in-law. After an early attempt, based on the bailiff's account of 1654–65, Windham introduced a new system of double entry book-keeping for the Green Book in 1673.[70]

From the outset, Windham believed that with investment and intervention rent levels could be maintained and failure averted. In 1673, he conceded a rent reduction to the tenants of Tuttington Hall Farm, from £47 to £37, but only as 'I was loathe to take the farme into my owne hands, not then living in the countye'.[71] When the tenant of Reepham Farm made a similar demand, William terminated the lease and took the farm in hand. He had already identified the farm for improvement, agreeing in 1668 to build a bakehouse and dairy.[72] In 1670 he also built 'a dairy at my house in Gresham' and leased the tenant 10 cows at £2 5s per head (our Type G). Diversification into livestock and dairy farming was clearly an early objective. He used Felbrigg Park and the demesne, with its extensive meadows, as a basis for this operation, receiving and holding stock before dispatching them to different farms as the dairies and bullock fattening units were established. In the Green Book, he kept 'an account of the stock of beast upon grounds in hand' and 'how the stock have been disposed to my own use'.[73] John Salman, the Felbrigg bailiff, co-ordinated the movement of stock and kept his own accounts, recording hundreds of transactions.[74]

The first sharefarming agreement concerns the improvement of neglected pasture in Felbrigg Park. In 1677, Windham agreed with Theophilus Waterson to plough up the Rush Close by farming it to halves (Type A). It was a detailed lease running for 13 years. In the first seven years, Waterson was to break up the pasture, level, drain, plough, muck and cultivate it. The rotation included 1. buck or summerley; 2. wheat; 3. barley; 4. oats to be followed by first years of 'olland' or fallow. To encourage Waterson, Windham accepted half the corn of the cultivated area, whereas he charged 18s an acre for remaining pasture. The terms were clearly set out and required no subsequent modification, indicating that it was a well known device for improving pasture. The lease successfully ran its course and shows that for a limited and easily controlled task, letting to halves worked well, but when combined with a new dairy farm and leasing a herd of cows it proved more problematic.

Between 1673 and 1677, William had kept Reepham Farm in hand, engaging Thomas Sexton, the grazier and tenant of Sustead and Metton meadows, to work the farm alongside his own fattening and butchering enterprise, and the dairy farm at Gresham. Salman supervised the operation and kept the records. Despite his efforts, William calculated in his bi-annual accounts, that he made a substantial loss in 1675 and 1677. His expectation that the sales of beef, corn and sheep would pay the rent and cover his costs proved optimistic. The problem was that his costs, including parish rates, tithe, purchase of stock, and Sexton's wages and husbandry charges were more or less fixed while prices declined. Moreover, Sexton's accounts required the closest scrutiny. To reduce the level of management Windham entered into a sharefarming agreement, as he said, 'the trouble of looking after this farm made me let it to halves'. In the agreement, for five years, Windham leased a dairy herd of 10 cows to Sexton at 45 shillings a head, (Type G) and the arable they 'plowed to halves' (Type A).[75] Windham provided the land, Sexton the labour, and they divided the corn crop at the end of the year. Windham calculated that the income from the cows and his half of the corn would equal the old rent of £52 a year. However, for three years Sexton failed to pay the full rent for the dairy, while Windham's return on the arable declined with the falling price of corn. He noted at the end of each account, 'Lost for want of a Tennant'.[76] Nevertheless, he did not abandon the idea, suggesting that other reasons contributed to its failure. Firstly, the agreement was poorly drafted with no clear division of responsibility, allowing Sexton to submit every kind of demand. Secondly, it made no provision for independent assessors to ensure that the Windhams received a genuine half. Thirdly, William may well have taken advantage of the clause which allowed him to graze additional stock 'for my own benefit', thereby reducing the pasture available for the cows. Finally, Sexton had no interest in the well-being of the dairy herd as Windham took responsibility for all herd replacements. Far from reducing the management burden, the arrangement required even closer scrutiny as so much depended on the honesty of the tenant. In technical terms, Windham had lost the initiative to the tenant, which left his working capital dangerously exposed. He made the situation worse by agreeing to a five year term, allowing the tenant to carry forward debt from year to year.

To correct these shortcomings, Windham engaged Edmund Britiffe, a modest but enterprising landowner from nearby Baconsthorpe.[77] His meticulously drafted agreements ran to several clauses and were designed to avoid disputes. Crucially, the stocking rate was clearly stipulated, Windham's share of the husbandry costs defined and Britiffe retained as arbitrator. In 1678 he arranged with John Masters the 'hiring a dairie of Cows' and 'plowing the ground to halfs' in Felbrigg Park .[78] Windham's motive was 'to have the conveniency of a Dairie near me, & bee free from the trouble of plowman'. Masters cultivated areas directed by Windham, while his wife ran the dairy. The scheme lasted until 1681 when Masters' wife died 'which made him not fit for imployment'. However, his successor, Thomas Cussens continued until 1693, leasing the dairy and 'sowing to halfes' 67½ acres in the Park with wheat, maslin, barley, buckwheat, peas, vetches, turnips and clover.[79] Less successful was the agreement with John Fincham for 'Selfe's Farm' at Roughton.[80] Windham made no charge for the building, laid on a small dairy herd and gave Fincham a loan to set up his farming operation. However, the operation collapsed in 1681 with Fincham 'too poor to continue', leaving Windham to write off debts and buy up his stock. This debacle, combined with Sexton's rising debts at Reepham Farm probably explains why Windham leased both farms at reduced rents in 1681, and only persevered with the Park Dairy, as it supplied the household.

Apart from the Park Dairy, Windham abandoned letting to halves in the Felbrigg area, preferring to take vacant farms in hand and develop his livestock enterprise with direct labour. A highly commercial business emerged, whereby he received stock from failed tenants, purchased store cattle from local fairs, fattened them on the Beef Closes at Sustead and sold them to butchers, in much the same way as his father had done in the 1620s. On the arable side, Salman paid men to carry out the husbandry and sold the corn to local millers. In this way Windham made a useful supplementary income and built up a network of local contacts. During the course of the 1680s, he persuaded several of these local butchers, grocers and small landowners to accept leases at reduced rents. By that time he was prepared to make concessions to attract well capitalized tenants. His decision illustrates the problems facing land-owners as they intervened in farming operations. Direct farming required intensive management, but was flexible and allowed him to keep control of his working capital, while letting to halves in theory reduced the level of monitoring, but put his working capital at risk. As we have seen, letting to halves worked well in a limited situation like Rush Close, or where strict commercial criteria did not apply as with the Park Dairy. Beyond that building up a network with local butchers, which led to their accept-ance of tenancies albeit at reduced rents, was a much safer option. Given the well developed markets for meat and grain in populous east Norfolk this strategy made sense.

Windham continued letting to halves on offlying properties as a way of attracting tenants and supporting rent levels. For example, at Scrivener's Farm, Dilham, which he 'could not let ... without great abatement', he agreed with John Applebye 'to sow Mack's Farm to halfes'.[81] He kept control by renewing the agreement annually, which is the classic arrangement. In 1683 he persuaded Applebye to take a lease at a reduced rent in 1683, with an abatement for the first year. Less satisfactory was the agreement in 1682 for part of Parke's Farm at Alby, 'for fear I should not git a Tennant I agree with Dan Shepherd to live in the house rent free'.[82] He also leased him eight cows at a reduced tariff and they sowed the ground to halves. Windham retained the right to terminate the agreement at three months' notice but in the event Shepherd refused to surrender his occupation and stayed for four years. In 1686 Windham bought him off with four cows and £2 worth of hay to avoid a legal suit. This appears to have been the last straw. From 1687, letting to halves was phased out, in favour of a uniform policy of rent reductions; the marginal recovery of corn prices in the mid-1680s making this strategy possible. However, when they slumped again in 1689, the year of Windham's death, his wife Katherine faced a clutch of vacant farms at Felbrigg, Reepham, Tuttington, Alby, Dilham, East Beckham and in Essex, which forced her to intervene.[83] For Parke's Farm at Alby, she concluded corn agreements with the two failed tenants, which continued into the early eighteenth century (Type J). With John Lound of Beckham Hall, whose debts amounted to £210, she devised a scheme to enable him to farm and pay off his debts.[84] She bought his cows, set the sum off against the debt, and leased them back to him (Type G). She made no charge for the land and buildings, received payments in kind and allowed substantial sums for marling, repairs and tax. With the rise in corn prices in the mid-1690s, she took the opportunity to let Beckham Hall and other farms at reduced rents. The dairies disappeared and the Windhams never again entered into working partnerships with their tenants. From that time they directed investment into fixed capital: land purchase, enclosure, buildings and marling. The Windhams continued to receive some rents in kind, including butter and cheese from Edmund Bale of Wymondham worth £20 a year.

Tables 5.1 to 5.4 summarize the performance of the Felbrigg estate between 1674 and 1696, and 1707 to 1717, and identify the role played by letting to halves and corn agreements as a way of dealing with vacant farms, and as an alternative to direct farming or reduced rents.[85] In Table 5.1, we can see its introduction in 1677 and phasing out in 1685, despite the evidence of more farms being taken in hand. Allowances included expenditure on repairs, tax, abatements and the buying-in of stock equipment, when tenants failed, as at Crownethorpe Farm in 1676. Windham wrote off the debt of £440, noting in the Green Book, 'I did looke as carefully after this farme as I could well doe and kept a strict account of it, that my son may see the inconveniency of having farmes come into his own hands'.[86] This debacle, combined with events at Reepham Farm, prompted the experimentation with letting to halves the following year. Table 5.2 showing the years 1688–96, shows the impact of the collapse of prices in 1689, with several more farms coming in hand. As far as we know, Katherine Windham did not resort to letting to halves, but leased cows and accepted rents paid in kind. Table 5.3, covering 1707–17, shows a fully let estate, but still with some rents paid in kind, and income supplemented by the sale of timber. Table 5.4 shows how rents were reduced to attract tenants and lease farms. The enhanced rent at East Beckham reflects the resumption of land purchase in the early eighteenth century. Although the documentation of letting to halves at Felbrigg provides the most detailed agreements, we can see from these tables that the extent of the practice was limited; this suggests that while Windham liked the idea, implementation was a different matter.

Why did sharefarming at Felbrigg prove to be of limited success? First, we need to remember the depth of the crises in the 1670s which left around a third of the estate in hand. The agreements for the dairy farms were introduced in the most adverse climate, when all other measures – fixed rents and direct labour – had either failed or proved unworkable. At least with cow leasing and farming the arable to halves, diversification into stock and dairy farming was achieved, if only in the short term and as a precursor to reducing rents. Clearly, the agreements, even those

Table 5.1 Summary of rental, arrears, receipts and allowances at Felbrigg, 1674–87 (£)

	1674	1675	1676	1677	1678	1679	1680	1681	1682	1683	1684	1685	1686	1687
Nominal rental	2,114	2,090	2,090	2,095	2,105	2,180	2,280	2,285	2,300	2,305	2,300	2,280	2,250	2,230
Old debts	440[*]	440[*]	440[*]	440[*]							715[*]			
	4,093	1,757	2,275	1,533	1,685	1,766	2,191	2,892	2,547	2,476	1,909	1,805	1,798	2,354
Park & lands in Felbrigg area[#]	179	253	245	179	268	329	263	272	333	316	471	440	447	282
Farms in hand	52			127						265	265	255	255	350
Farms let to halves				62	62	100	100	100	216	216	74	10	10	10
Rent charge	1,883	1,837	1,845	1,727	1,875	1,761	1,927	1,913	1,847	1,498	1,490	1,575	1,537	1,498
Total due from arrears & rent	5,976	3,594	4,120	3,260	3,560	3,527	4,118	4,805	4,394	3,974	3,399	3,380	3,335	3,852
Gross receipts	4,117	1,200	2,661	1,680	1,818	1,438	1,313	2,226	1,847	1,266	1,580	1,524	1,034	1,577
Net receipts	3,280	1,058	1,765	1,569	1,533	1,096	943	1,895	1,609	1,113	1,411	1,217	774	1,212
Allowed	837	142	896	111	285	340	370	331	238	153	169	307	260	365

[*] Debts written off   [#] Estimated values of closes taken in hand or farmed with the park; dairying, stock fattening, and tree planting

Source: E.M. Griffiths (ed.) *William Windham's Green Book 1673–1688*, Norfolk Record Society, vol. LXVI (2002).

Table 5.2   Summary of rental, arrears, receipts and allowances at Felbrigg, 1688–96 (£)

	1688	1689	1690	1691	1692	1693	1694	1695	1696
Rental	2,228	1,780	1,600	1,594	1,566	1,543	1,506	1,494	½yr
In hand	776	389	489	582	614	607	565	566	565
Corn rent			60	60	60	60	60		
Rent of stock			47	37	103	49		97	63
Rent charge	1,462	1,391	1,051	952	892	876	921	928	453
Arrears	2,638	2,633	2,490	2,974	2,533	2,523	1,074	955	1063
Total due	4,100	4,024	3,588	3,963	3,528	3,448	1,995	1,959	1,580
Gross receipts	1,466	1,534	614	1,428	1,004	1,243	1,040	896	444
Paid in cash	1,278	654	388	603	586	729	644	540	217
Paid in corn	102	98	104	276	112	110	163	59	199
Allowances	86	782	122	549	306	404	233	297	28
Debt written off							1,130		1,096

Source: E.M. Griffiths, 'The management of two east Norfolk estates in the seventeenth century: Blickling and Felbrigg', (Unpublished PhD thesis, University of East Anglia, 1987), p. 396.

drafted by Britiffe, left the landowner's working capital vulnerable, but the alternative of tenants failing repeatedly proved even more costly.[87] However, as we shall see at Raynham and Blickling, there were ways of exerting tighter controls and preventing the initiative passing to the tenant. Perhaps the Windhams showed some naivety in the implementation of the practice. When they limited their objectives, as with the lease for Rush Close, and the recruitment of a tenant for Scrivener's Farm at Dilham, the device worked well and achieved the desired results.

The Windhams' naivety raises another question – their suitability for administering the practice in person. What is obvious from William Windham's comments in the Green Book is the rising tension between him and his failing tenants, and the anguish he experienced trying to balance benevolence with hard economic necessity.[88] Despite his best efforts, engaging with tenants at such an intimate and domestic level offended his sensibilities and proved increasingly distasteful. The problem for the Windhams was as much social and cultural as economic and administrative. This observation reminds us of the point made by Hilton, that the main obstacle to success was the social distance between the parties which made deception easy and co-operation difficult.[89] The examples from the early seventeenth century appear to contradict this view, but the economic climate in the 1630s and 1640s was very different

Table 5.3   Summary of rental, arrears, receipts and allowances at Felbrigg, 1707–17 (£)

	1707	1708	1709	1710	1711	1712	1713	1714	1715	1716	1717
Rental	1,300			1,450							
Rent charge	1,291	1,286	1,334	1,448	1,434	1,391	1,350	1,400	1,380	1,384	1,338
Arrears	939	530	563	600	738	264	199	281	242	392	53
Total due	2,230	1,829	1,900	2,067	2,182	1,665	1,549	1,681	1,621	1,776	1,391
Gross receipts	1,700	1,266	1,300	1,329	1,919	1,466	1,268	1,439	1,229	1,723	1,250
Paid in cash	1,097	910	865	732	893	822	772	802	780	1189	734
Paid in kind corn/meat/dairy	196	70	146	225	573	197	142	105	136	100	83
Net receipts	1,293	980	1,011	957	1,466	1,019	914	907	916	1,289	817
Allowances	407	286	289	372	452	447	354	532	313	434	433
Timber sales	557	250	192	200	620						

Source: E.M. Griffiths, 'The management of two east Norfolk estates in the seventeenth century: Blickling and Felbrigg', (Unpublished PhD thesis, University of East Anglia, 1987), p. 467.

Table 5.4 The movement of rents at Felbrigg, 1673–1717 (£)

	1673	*1681*	*1687*	*1692*	*1695*	*1709*	*1717*	
Felbrigg	584	399	356		431	680	673	
In hand/let to	175	359	357	331	191			
halves	759	758	713		622	680	673	–11%
Tuttington area	263	272	227		146	148	136	
In hand			39	39				
Corn rent					60			
Dilham	81	82	47			82	82	
In hand			30	77	77			
East Beckham		204	204	180	112	254	275	
In hand/let to				24	86			
halves								
Reepham						41	41	–21%
In hand/let to	52	52	52	50	50			
halves								
Wicklewood &	300	301	301		241	245	245	–18%
Crownthorpe								
Suffolk	230	217						
In hand			203					
Essex	429	412	412					
In hand								
Farm rents	1,887	1,887	1,523	892	928	1,450	1,453	
In hand	227	413	706	674	566			
Total	2,114	2,300	2,230	1,556	1,494	1,450	1,453	–11%

Source: E.M. Griffiths, 'The management of two east Norfolk estates in the seventeenth century: Blickling and Felbrigg', (Unpublished PhD thesis, University of East Anglia, 1987), pp. 477–80.

from the 1670s and 1680s, when costs needed to be firmly controlled. The experience at Blickling and Raynham suggests that forms of share-farming, or unconventional agreements, were more successful when implemented by a professional, preferably from a similar social background to the tenant, or by outsiders, such as trustees, who fully understood the working of the system, offered a degree of detachment, and were paid to be hard nosed when the occasion demanded.

At Blickling Sir John Hobart's interventionist strategy was implemented by a team of professionals, and illustrates the wisdom of employing intermediaries to conduct estate business. They too supported tenants and invested heavily, but from the outset they exerted much tighter controls over expenditure and maintained a more even relationship with the tenantry. They were also acting under greater constraints. Sir John's team, led by Robert Dey and John Brewster did not possess the same freedom to experiment and lose money. Sir John's indebtedness, in fact, left very little room for manoeuvre. Brewster could not risk half the estate falling in hand; he needed to anticipate rather than react to disaster and to control costs at all times. Dey, the family lawyer, retained a guiding hand until 1671, but from 1669 Brewster initiated estate management policy.[90]

The Blickling estate had different origins from Felbrigg, Raynham or Hunstanton. It was not an ancient estate, but one created by Sir Henry Hobart between 1597 and 1625. As we have seen his policy was to cherry-pick small estates as they came on the market, invariably properties with well organized demesne lands and scope for improvement. The only estate that contravened this principle was Blickling itself, the prestigious home of the Boleyn family. The properties bought for investment included, on the heavier lands to the south and west of Norwich, Intwood Hall, Langley Abbey and Grange, and the Parks at Wymondham. To the north of the city, on the heathlands, he acquired Horsham St. Faiths, Hevingham and Saxthorpe. During the first half of the seventeenth century, these holdings were improved and leased successfully to large tenants. However, in the 1660s, the tenants started to fail, forcing the Hobarts to intervene.

The earliest evidence of the Hobarts' strategy is a plan for Langley Abbey in 1666, devised for the 'Aby to answer a profit in lieu of rent'. Dey and Brewster created three separate enterprises. First, they leased the tithes for £60; secondly, 127 acres with 2 tenements were allocated to maintain a dairy of 30 cows, leased at 45 shillings per head, which they estimated to be worth £91.50; and thirdly, 258.75 acres were to be cultivated, for an expected return of £78.50, of which 90 acres were kept for mowing, 'keeping horses for ye tillage', young cattle or what else a good husband should think.' Already there was a 'full stock of cattle upon ye ground and 12 beast at stake with Turnepps'. The 169 acres in tillage were divided into 30 acres of winter corn, 18 acres of peas, 33 acres of oats, 52 acres of barley, 16 acres of turnips and 20 acres of summerley, confirming the use of turnips in rotation by 1666. No accounts survive for the implementation of the plan at Langley Abbey; the arrangement lasted for three years, ending in 1669, when the holding was leased, with a 10 per cent reduction in rent, abatements for the first and final year, and Sir John assuming responsibility for repairs. We do not know how the arable was cultivated, but most likely by contractors which Brewster used elsewhere.

In 1666, a similar scheme was implemented at Langley Grange. Henry Gallant's accounts show that in 1668 the dairy was let to Henry Nott and his wife at 40 shillings a cow and Robert Horne cultivated the arable; he did not sow the land to halves, but agreed contract rates for the job.[91] We note that Brewster kept the dairy and arable entirely separate, avoiding the reliance on a single tenant. In 1675, Horne, having built up sufficient capital, took over the tenancy and purchased the dairy of 32 cows. In 1668, when Little Park Farm at Wymondham failed, Brewster leased the dairy of 25 cows at 55 shillings a piece to Goodman Wakefield, blacksmith, and his wife Alice, who featured prominently in his accounts paying the rent and negotiating abatements. The arable land was farmed by John Knight and John Cullyer on a contractual basis. In 1680, Little Park was re-let at a rent reduced from £220 to £160 a year, while John Knight moved to Great Park Farm and took on the tenancy at a similarly reduced rent in 1687. In 1677 John Cullyer had moved on to lease the dairy and farm the arable at Intwood Hall; he worked the farm until his son Richard accepted the tenancy in 1681. These agreements show how contractors, performed various tasks, moved from farm to farm and built up sufficient capital to take on fixed rent tenancies, all the time negotiating hard for concessions, abatements and reductions in rent.

Brewster's management of Intwood Hall between 1677 and 1680, demonstrates the precise calculations involved if such complex arrangements were to be successful. In the agreement of 1677, John Cullyer leased 20 cows at 48 shillings a piece and paid the rent for the cows by farming the arable on a contract basis. Brewster specified the rates to be paid for ploughing, harrowing, sowing, reaping, carrying and laying corn into the barn; also for cultivating turnips, making hay, carrying and spreading muck, and for looking after cattle. The total cost of husbandry was valued at £22.50. Brewster added to this figure allowances for repairs to buildings and the use of Cullyer's implements; the total sum was then set against Cullyer's rent of £48 for the dairy. Brewster reserved the right to review the terms annually, which meant adjusting the farming rates and putting on more cows, if the cost of farming the arable exceeded the rent for the cows. This happened after the first year when Brewster made a small loss, but in the final two years, with five more cows and slightly reduced rates he made a reasonable profit, and Sir John Hobart still benefited from the sale of the whole corn crop. In this arrangement Brewster kept the initiative. The tenant knew that if his farming costs were too high, Brewster would shift the emphasis to dairy farming. No sharing of the corn crop reduced the opportunity for fraud, and left Cullyer with the incentive of making a profit on his cows. Such a scheme required meticulous recording; Brewster's man on the ground Francis Eagle filled pages with minute detail, which all had to be checked by Brewster. When Cullyer's son agreed to seal a lease, despite a 14 per cent reduction in rent, Brewster terminated the scheme. From his perspective, the objective of securing an able and well capitalized tenant had been achieved.

Unlike Windham, John Brewster never tried to maintain rents at artificial levels. From the outset he favoured negotiated concessions supported by written leases often containing directions to improve husbandry, notably the use of turnips on the light soiled farms. The lease for Abbey Farm, Horsham St. Faith's in 1666, one of a series designed to improve this difficult farm, included the most detailed instructions for the cultivation of turnips.[92] Turnips had, in fact, been used at Abbey Farm prior to 1666 – the sale of stock, corn and equipment by the outgoing tenant in that year included a crop of turnips – the purpose of the leases was to regularize and extend their use. Despite this intervention, and a similar lease in 1673, difficulties continued. In 1679, when prices plummeted, Brewster concluded a Corn Agreement guaranteeing the tenant a market and price for his corn. This was not sharefarming, but a straightforward subsidy by the landowner (Type J). While sharing the risk of the markets, subsidies were more easily controlled than sharefarming with tenants, while the proximity of Horsham St. Faiths to Norwich provided ready outlets for the grain. The net return at Abbey Farm, at 62.5 per cent of the rent charge, remained the lowest on the estate, but better than the 50 per cent recorded in the 1650s and early 1660s. More to the point, with the subsidy, tenants at St. Faith's survived and improvements were implemented; for William Wix at nearby Horsford, Brewster noted 'a dairy to be made'.

As at Felbrigg, the nature of the intervention on the Blickling estate changed noticeably after 1682 as prices staged a modest recovery. The priority was to find tenants with working capital and the right kind of expertise to operate in an increasingly difficult market. Investment was directed to attracting and supporting these men. At the same time the exploitation of the estate at Blickling was intensified with new farms carved out of the park and heathland, and neighbouring farms drawn into a network to supply the household. Brewster's accounts reveal a complex local economy with local tenants supplying a range of goods and services. Thomas Allen at Saxthorpe provided oats, hay and barley for brewing; William Trappet at Blickling, kept the gardens, including the greenhouses, ran the dairy and took on several of the new tenancies at different times. New tenants, living near the hall, often combined a variety of tasks. William Smyth at 'the Flash' collected rents, while his wife supplied milk, cheese, honey, peas, pears and apples. Goodwife Bongen kept pigs and fowls, while her husband obtained fish, plaice and soles, herrings and crabs. Goodwife Springall gathered 'hearbes for dyet drinks' and Mrs Jell supplied oatmeal. Richard Gay of 'the Park Grounds' reared bullocks, including ten steers at turnips and provided butchered meat, while Matthew Fairchild, the blacksmith, undertook ironwork and all sorts of repairs. Tenants also served as carriers, taking fresh food to the Hobarts' house in Norwich, surplus corn to Coltishall, and bringing back wine, luxuries, coal and other household items to Blickling. In effect, Brewster instigated a system of guaranteed prices and markets to support tenants, with the added benefit of curtailing Sir John Hobart's expenditure.

Blickling differed from Felbrigg in that sharefarming was never adopted as an estate policy. The Hobarts intervened to support tenants with working capital, subsidies and guaranteed markets, but Brewster studiously avoided schemes which involved sharing inputs and outputs with tenants. However, that does not mean sharefarming did not exist on the Blickling estate. A reference to farming to halves appears in the documents supporting the sale of the Langley estate to the tenant Richard Berney in 1718. In the survey of 'the Langley Estates belonging to Sir John Hobart and ye manner of ye occupines and how they are parcelled out by Mr Berney', written against,'the Clover lands hired by Mr. Church', is an insertion in the tiniest writing, 'These lands when in tilth were halfed wth Mr Berney and were laid down with clover at Mr Berney's charge and are now let for the two years'.[93] Brewster's estate accounts make no mention of the practice and simply record that Richard Berney paid a rent of £474 for Langley.[94] The likelihood is that on these large holdings, farming to halves occurred as a matter of course between tenant and sub-tenant, but was not considered appropriate between landowner and tenant. As a small landowner in his own right, Brewster would have been fully aware of the pitfalls, which Windham discovered so painfully at Felbrigg, and was sufficiently astute to work out other less risky ways of supporting tenants, particularly where local markets offered so many opportunities. The impact of Brewster's policy can be seen in Table 5.5 which shows a substantial increase in rents at Blickling, where farms and enterprises were developed, combined with sharp reductions of rents on offlying properties to attract tenants.[95]

Table 5.5 The movement of rent on the Blickling Estate, 1665–1703 (£)

	1665	*1686*	*1701*	*1703*	
Blickling, Hanworth Tithes	309	c.450	514	534	+73%
Chapelfield House, Norwich				67	
Intwood, Keswick, Swardeston	503	c.430	434	435	−13%
Langley Abbey and Grange	494	450	460	sold	−7%
Wymondham Great and Little Park	550	423	433	433	−21%
Wood Dalling	200			209	
Horsham St. Faiths	438	374	sold		
Hevingham	365	310	sold		
Saxthorpe	216.50	160	sold		
Total rent	3,075.50	c.2,797	2,041	1,678	

Source: E.M. Griffiths, 'The management of two east Norfolk estates in the seventeenth century: Blickling and Felbrigg', (Unpublished PhD thesis, University of East Anglia, 1987), pp. 401, 409, 430, 440.

John Brewster's astute management at Blickling shows how yeomen farmers were familiar with all sorts of strategies for turning a profit and making a living from the land. With his 200 acre farm at Fundenall, near Wymondham, he had much in common with Thomas Jackson of Budworth and Edmund Britiffe of Hunworth, who owned land, let to halves and supplemented their income by offering their services to gentry landowners. As we have seen, the Britiffes, in particular, profited from marketing their professional skills and expertise in agriculture. By the 1660s, known as yeomen in the first half of the century, they had established themselves as parish gentry, occupying comfortable manor houses at Cley, Hunworth, Stody and Baconsthorpe.[96] Edmund Britiffe, who drew up the letting to halves agreements for William Windham, also enjoyed a close relationship with Sir Charles Harbord of Gunton, for whom he drew up leases and agreements, arranged loans and mortgages, collected interest, conducted correspondence and acted as intermediary in disputes, notably between his brother, Simon and Sir Charles. In a letter dated 1666, Simon Britiffe informed Sir Charles, his long suffering father-in-law, 'I am now setting my estate [which was in my own hands] to halfes with my tenant Flaxman'.[97] Clearly, farming to halves was common in the Britiffes' social circle, and Edmund had drawn up formal agreements for William Windham based on his own experience.

While Edmund Britiffe was influential in his own locality, his son Robert became a man of considerable weight in the county.[98] Trained in the law, he served as steward, family lawyer and adviser to leading families, including the Townshends and Walpoles. As Sir Robert Walpole's political agent, he served as Recorder and MP for the city of Norwich from 1714, and was famously a shrewd operator in the land market, negotiating purchases for Walpole in the 'indestructible fields of Norfolk'.[99] At the same time, he amassed for himself a large scattered estate across Norfolk and Suffolk and secured advantageous marriages for his two daughters to the heirs of Blickling and Gunton.[100] In the early part of his career, Britiffe had served as steward to the Berneys of Westwick. A letting to halves agreement was found amongst the estate papers, between John Berney and John Ollyet, yeoman from 1698.[101] It resembled one of those in Robert Loder's account, in that it involved the farming of a widow's holding: Ollyet paid rent for one end of the house with buildings, gardens and two closes, and sowed the 'rest of ye lands' to halves, according to the rotation laid down by Berney. There is no proof that Britiffe was the author, it may simply be a further example of the practice in the Broadland area. However, in the marriage settlements he drew up for his daughters, he listed the holdings to be settled on them, referring to the tenants by name and their under-tenants, for example, 'all that farme in Blickling now or late in the use of W. Smith or his assignees, under-tenant or under-tenants'.[102] This format indicates his familiarity with agreements below those that existed between landlord and tenant. Britiffe also negotiated the sale of the Langley estate in 1718, where we found the reference to farming to halves. Given this background and later connections with the Townshends, we thought Robert Britiffe might have been responsible for the adoption of letting to halves at Raynham in the 1690s. This was not the case. For as we have seen, forms of sharefarming were as common in west Norfolk as in east Norfolk.

The west Norfolk examples of letting to halves, in the second half of the seventeenth century, are drawn principally from the archives of the Townshend family. However, material from Houghton and Hunstanton confirm that forms of sharefarming continued to be practised elsewhere. The Houghton papers are frustratingly sparse, but nevertheless revealing. The evidence for letting to halves appears in a chaotic book of receipts. One entry starts, 'what is delivered to Will Rowardes halfeing of ye plowground for seed and other things beginning Mich. 1689'. Another, more detailed and conclusive, is a corn account of 1692, which shows the corn crop to have been shared between Walpole and his tenant Christopher Rodwell, 'rec'd of my part ... of 6 acres of wheat ... of 4 acres of Rye ... of 20 acres of pease and fetches [sic] ', and 'Memo: Rodwell tooke out of this crop all my part of seed for wheat, rye and pease'.[103] Another document, a book of 'small rents' from 1647–1681, sheds light on the letting of 'cow grasses' to smallholders, a practice which also appears at Hunstanton and Raynham, but not in east Norfolk. Cow grass was the summer grazing leased by men – and women – who enjoyed grazing rights on the winter stubbles.[104] The rental lists 'cow grasses' let annually to about a dozen tenants, each renting one and half to three acres at 2s per acre, suggesting a stocking rate of a cow to an acre and a half. Invariably, they also rented a few acres of arable, and performed other tasks, such as supplying food for the household, carrying goods and general labouring work. This arrangement indicates the presence of cowkeepers and smallholders earning their living from a variety of sources.

At Hunstanton, the accounts kept by the trustees for Sir Nicholas Le Strange from 1669 to 1683 show a similar arrangement with pastures, grasses and cattle leased to tenants as a way of securing some return from vacant holdings.[105] For example, 'Allen Collen his corne farme and money farme un-let – he to pay at Mich next £3 for the grass of part of the lands'. When the tenant of Hunstanton Meeles (sand dunes) failed in 1672 it was 'let to the townsmen' with 13 men and two women renting pasture for 30 cows at 3s 4d each; in 1674, 26 tenants paid 2s per horse and 6d per cow for 72 cows, and from 1675 they paid a fixed sum of £3 for 'ye meels'. We remember this holding was cultivated to halves by his grandfather, Sir Nicholas Le Strange in 1637.[106] Several foldcourses and flocks, as well as farms, were divided and let in halves, and even thirds. Mr Travies of Sedgeford Hall leased '½ ye North Flock, ½ Hall Botome Close, ½ old Mill meadows ...' Nicholas Le Strange held the other half of the Sedgeford flock and contributed 203 sheep 'towards the half stocke for this flock Mich 76–77... Mr Richard Travies is to lay the other halfe and to pay for that moiety £10 for a years rent.' Francis Grippe leased 7 acres in Ringstead and a '3rd pt of Ringstead Foldcourse', while Richard and John Harrison rented 'the other 3rds'. The division of flocks and foldcourses (Type E) was a continuation of the practice first identified in the early seventeenth century, but there is no evidence of the Le Stranges farming the arable to halves as they did then. The documentation for this period is far less detailed: there is nothing to compare with Alice's accounts or the notebooks of Sir Nicholas.

These examples from Houghton and Hunstanton, when compared to those from Felbrigg and Blickling, indicate a different type of rural economy and slightly different reasons for farming to halves. In east Norfolk, with its fertile soils, intensive agriculture, dense population and well developed market for food, farming was less precarious than on the light sandy soils of west Norfolk. For landowners in East Norfolk sharefarming was an option rather than a necessity. After experimenting with letting to halves, Windham was soon able to revert to a system of relying on local butchers, as his father had done, while Brewster skilfully continued to manipulate contract rates, prices and local markets. In the first half of the seventeenth century, no evidence of farming to halves was found at either Felbrigg or Blickling. In west Norfolk, conditions were different. Much of the region consisted of marginal land, including marsh and breckland, which remained undeveloped or was in the process of being brought into cultivation. It was in these situations, which entailed considerable risk to landowners and tenants, that the Le Stranges and Townshends farmed to halves. Few tenants possessed sufficient capital to stock large farms, or to bear the risk of taking on a large tenancy at a fixed rent, so landowners were often forced to lease flocks and foldcourses to syndicates and farm the arable to halves. Without landlord assistance on a substantial scale, and buoyant grain prices, these holdings were simply not viable. The situation became progressively worse as prices fell in the 1660s, necessitating further subdivision of holdings, flocks and generous agreements to support tenants.

However, for smallholders and landless labourers, the extensive farming regime in west Norfolk, with ready access to waste and commons, provided all sorts of opportunities. Rights of grazing on the winter stubbles allowed them to keep livestock and eke out a living as cowkeepers, cheese and butter makers. These flexible and low cost enterprises could be expanded at will by subletting, pooling resources, establishing partnerships, hiring cow grasses and farming to halves. At Hunstanton, and along the north Norfolk coast at Stiffkey and Morston, farming was often combined with fishing and fowling, creating economically dynamic and diverse communities.[107] In the 1670s and 1680s, as grain prices continued to fall, these smallholders took advantage of the vulnerability of large farms; we find them snapping up portions, renting pastures, sharing flocks and setting up dairies. The fragments of evidence from Houghton and Hunstanton provide a glimpse of this process. The Townshend papers facilitate a fuller and more contextualized picture of how sharefarming was used by landowners and smallholders in west Norfolk during this difficult time.

Sharefarming on the Raynham estate, as we already know, dated back to the fifteenth century and existed in the first half of the seventeenth century. The remainder of this case study will show how the Townshends, or more accurately their stewards, continued to farm to halves at Raynham until the early eighteenth century, when it disappears from the estate records. Picking up the threads from the 1660s, the focus will be firstly, on the period from the early 1670s to 1687 when Horatio, 1st Viscount Townshend and his stewards, Timothy Felton and Thomas Ward, revamped the management of the estate; secondly, from 1687 to 1697, when the estate passed into the hands of trustees for the 2nd Viscount, Charles Townshend, later known as 'Turnip Townshend'; and finally, from 1698, when Charles Townshend took control of the estate and terminated the letting to halves agreements instigated by the trustees.

In the early 1670s, Horatio Townshend, like his brother-in-law William Windham, embarked on a radical reform of his estate business.[108] Showing equal commitment and zeal, he drew up instructions for his bailiffs and auditors, scrupulously examined his accounts, and wrote up his observations. When absent in London, he conducted a vigorous correspondence with Timothy Felton and Thomas Ward. Following Raynham tradition, Townshend did not keep a central account book; as before, the six bailiwicks making up the Norfolk estate were accounted for separately. Their accounts detail the structure, tenurial arrangements and the succession of holdings, making it possible to create profiles of farms and trace the development of management policy over time. Farmalls provide an overview of estate structure at intervals between 1662 and 1701.[109] As we have seen, the six bailiwicks of the Townshends' Norfolk estate stretched across west Norfolk to the north Norfolk coast from the Raynhams, the Rudhams, Stibbard and Ryburgh, South Creake to the estate at Stiffkey, Langham and Morston. In 1681, Sir Horatio purchased William Ruding's estate at Toftrees, a few miles east of Raynham, with the proceeds from the sale of the Stanhoe estate. Each property possessed different characteristics and their management reflected this: even the letting to halves arrangements vary in their detail as they were adapted to different circumstances and farming regimes. Apart from Stibbard and Ryburgh, where no evidence was found, farming to halves occurred on every part of the estate between the 1670s and 1690s.

The lands at East, West and South Raynham, the core of the medieval estate, do not offer the most detailed evidence of farming to halves, but as at Felbrigg and Blickling, flexible and informal agreements, including farming to halves, were adopted and often associated with provisioning the household. Known for its poor light soils the Raynham lands, which lie south of the river Wensum, also included extensive areas of meadow and pasture, ripe for improvement.[110] The 1662 farmall refers to brecks and foldcourses leased for sheep feed amongst the park lands near the New Hall at East Raynham, but by the early 1670s tenancies had been created with evenly sized enclosures, homestalls and barns.[111] By 1684 the value of the lands in the park and East Raynham had increased from £542 to £866 a year, an improvement linked primarily to the expansion of livestock and dairy farming. As at Felbrigg, diversification into dairying was not easy. It required specialist buildings, skilled labour and tenants with sufficient capital to purchase stock and pay the rent. Like William Windham, Horatio needed to be flexible and offer inducements; it is in this context, that discussion of informal agreements appears in the correspondence. For example, in 1676, 'For my Lord's husbandry without the parke and the wintering of his cows', Townshend agreed to pay the tenant an extra five shillings a cow for winter fothering with oat, barley and wheat straw until calving time.[112] When the tenant failed in 1679, Felton resorted to direct labour, using servants to perform the tasks. He then proposed a scheme for breaking up the holding, but subdivision required another barn costing £4 which Horatio vetoed. Nevertheless, as he was 'resolved to keepe nether cow, no hoge, dairy made or cow fellow', they leased a dairy of 21 cows and a bull to Francis Page for £50 a year, or 45 shillings a head, with additional pasture for £24. This agreement, reminiscent of the lease for the Park Dairy at Felbrigg, lasted only a season, as Page could not pay his debts and his wife was 'not able to carry on the Dary'.[113] Felton tried to persuade the wife of Peter Stringer, formerly the dairy maid to take it on with her husband farming the arable, but to no avail. As a temporary measure, Felton reduced the herd to ten cows and hired Page's dairy maid, who made good butter and cheese, at 3s a week. His plan to fatten off the rest of the cows backfired as they refused to eat the hay grown on the meadows, 'it being flaggy and course ... And the closes were not fit to be in tilth'.[114] To improve the pasture, Felton recommended to Townshend that the lands be put out to halves, explaining that it was a more efficient way of cultivating the meadows than using direct labour, 'it would quit costs to bring in a team of horses ... it will be done with less charge by putting it to halves'.[115] This was the method used successfully by Windham for Rush Close in 1677, but we do not know if Townshend acted on the advice. In 1682, with no substantial tenant forthcoming, Felton divided the holding in two, but not before telling Townshend that he had alienated tenants by keeping too many deer in the park. In the agreements, there is no mention of a dairy, but one tenant paid rent for eight cow grasses. The future of the Raynham Hall dairy, which supplied the household was secured in 1685, when Townshend noted in his memo book, with obvious relief, 'Bette Stringer entered the dayrey at Raynham'.[116] It was subsequently, and successfully leased by the trustees from 1687 for £50 a year.[117]

The experience at East Raynham shows a new flexibility, a recognition of the need to split holdings, an increased reliance on small-scale farmers, usually women, to run the dairies and probably a growing acceptance of subletting. Leases invariably included a restriction against subletting 'without special licence or consent from Lord Townshend in wryting', but there is every indication that tenants themselves resorted to such measures to secure skilled labour.[118] This can be seen more clearly at West and South Raynham. In West Raynham, the 1662 farmall records '83 acres in the neats pasture ... in the use of divers tenants at four shilling and sixpence per cow' – the same arrangement of cow grasses that we saw at Houghton and Hunstanton.[119] In 1674, William Carr leased eighteen cow grasses at 6s 9d per cow with his farm, while 'divers persons' rented the remainder.[120] A list of cow grasses, 'sett out 12 May 1683', shows 17 tenants leasing summer grass for 54 cows from May to November. The sale inventory of William Mitchell 'senior' of Weasenham – whose son leased the 18 cow grasses in 1684 – shows they were sometimes sublet. It records that he owned four cows, three calves, four swine and four horses. Set against his debt, he was allowed a sum of money for his manure and hay, the 'lands [he] let to sowe' to five men, and the cow grasses he let to four men at 6s 9d each. In other words, the older man rented cow grasses from his son, which he then sublet to his neighbours, who also sowed his arable to halves, for example 'to old Semper 1ac 1r at 5s 6d'.[121] At South Raynham in 'a note of leases and agreements made between Lord Townshend and several tenants in September and October 1682', we find, 'William Downhams bill of sale for £121. 7s 1d. And his agreement for his farm for one yeare at ... in money and halfe the corne crop'.[122] There is no sign of this arrangement in the accounts or 'farmalls'. Thomas Arnold leased the farm from 1684 and when he left in 1690, John Raby paid £14 rent for a dairy of six cows and cultivated the arable.[123] The following year, William Raby took a lease of the farm, with no mention of the dairy. Most likely it survived as part of the Raby family enterprise but simply disappeared from the estate documents.

The picture which emerges from the Raynhams is one of fluidity and diversity. From these scraps of evidence, just as at Houghton and Hunstanton, a community of smallholders and cowkeepers can be reconstructed, often working alongside large farmers, able and willing to take on holdings when the opportunities arose and the inducements were sufficiently attractive. As farms fell vacant, dairies were set up with men and women renting a few cows, to pay for the cost of cultivating the arable. Sometimes, the steward used direct labour, sometimes the tenant agreed to sow the arable to halves. At other times they agreed to accept a lease, otherwise the holdings had to be divided to attract a tenant. When a substantial tenant could be secured, holdings were often reassembled. For example, the farm at Patchley let in 1662 for £141 a year, was in 1683 divided between five tenants. In 1690 it was recreated and leased to a single tenant as 'Patchley House & Farmes & Meadows & Lands below the park'.[124] As the title suggests, this does not exclude the possibility of subletting. We can see that the activity at Raynham was principally driven by the demand of the household for goods and services, which effectively sustained a micro-economy, as at Blickling and Felbrigg. The Raynham household accounts minutely record hundreds of transactions, buying produce from smallholders, selling – or more likely lending them – cows and dairy equipment, and hiring their labour. These arrangements fostered all kinds of short-term, informal agreements. The 'Raynham effect' extended to the wider estate, providing a ready market for wool, corn, meat and stock for the new dairies established at Stiffkey, Langham and Toftrees. It also encouraged large and enterprising tenants, notably the Stringers and the Rabys, to implement new initiatives.[125]

The most sustained and explicit use of farming to halves was found at Stiffkey and the Rudhams, where the most deep-seated problems on the Raynham estate were encountered. The Stiffkey estate with lands adjoining Langham and Morston, formed the jointure of Horatio's mother Lady Westmoreland, and for long periods the family occupied Stiffkey Hall while the new hall was being built at Raynham.[126] The changing structure of landholding is shown in Table 5.6. From 1670, when it returned to the main estate, Townshend leased the property to tenants, with little success. The bailiff, Phillip Tubbing noted that he 'never rec'd any part' of Captain Rokewood's rent of £435, and in 1677 Mrs Abigail Mott surrendered her lease – at the lower rent of £375 – after just a year.[127] Finding a solution for Stiffkey was not an easy matter. The mansion, built by Sir Nicholas Bacon for his son Nathaniel, merited a genteel tenant, yet the estate required a practical man to farm the holdings and maintain the buildings and gardens. The decision to farm to halves may have been prompted by Townshend's desire to retain greater control and prevent further deterioration. As we have seen, Simon

Table 5.6   The changing structure of holdings at Stiffkey, 1670–1730

	Number of holdings							
Farm rents	1670	1677	1683	1684	1693	1701	1720	1730
<£1	2		2	2	3	3	2	2
£1 – £5	4	5	6	6	4	4	2	2
£5 – £20					1	4	2	1
£20 – £50	3	3	4	2	2	3		
£50 – £100						1	2	2
£101 – £200			1(1)					
£201 – £300						1	1	1
£301 – £500	1	1		1	(1)			
Holdings let	10	9	13	11	10	16	9	8
In hand or let to halves			(1)		(1)			
Total rent £	522.00	462.00	240.50	468.55	83.50	430.90	420.10	424.60
In hand or let to halves			rest to halves		89.25			

Source: Townshend MSS, RAS/A1/6; Estate Accounts, 1670–1730, RAD/B2–B3; RAS/A2/1–8; RAS/A3/1–8; RAD/C3–C5.

Britiffe had 'put to halfes' his estate at nearby Cley Hall in 1666. On the other hand, Townshend may simply have been unable to attract a tenant on any other terms. Whatever the motive, he concluded a letting to halves agreement with Richard Teasdale, a wealthy Stiffkey butcher in 1677. Unusually, it is set out as a formal lease, recording in minute detail the working of the arrangement.[128]

The agreement, signed by Tim Felton, Philip Tubbing and Peter Stringer, was for seven years. Teasdale paid £122.37 rent for the hall, foldcourses and 116 acres of meadow and pasture. Townshend stocked the foldcourses; Teasdale provided security, in the form of a renewable bill of sale on his estate and personal goods, for the payment of rent. Valued at £340.67, his stock comprised 960 sheep, 20 cows, 16 horses and 20 swine.[129] He also agreed to maintain the garden, orchard, walks, fruit trees and other plants and to observe the covenants to promote good husbandry, while Townshend allowed him to sow turnips in the arable closes 'for his own use'. The arable land of 315 acres, Teasdale sowed to halves. He agreed 'to plow, summertill and sowe to halves at his own cost and charge all said demised arable lands of Lord Townshend … to weed, keep, mow, cut, carry and inn all the said corn into the barn, and thrash and put up all the said corn and grain', and to deliver into

Townshend's granaries at Stiffkey 'one whole moiety or half parte of all the said corne or graine well dressed and delivered out'. The parties to the agreement chose four able persons to 'estimate a prise and judge the quantite and qualities of the corn …' These independent 'apprisers' assessed the yield of the standing corn in August and produced a written estimate; in 1680 it listed the acreage sown, the yield per acre, and the total expected. From the total, less tithe, it left the 'rest to be divided'.[130]

The estimate formed the basis of the accounts, written on small loose sheets of paper, often inserted in letters, *ad hoc* in format and very difficult to follow. The 'charge' included Teasdale's rent and the anticipated sales of corn, hay and grass. The profits comprised the actual sales, rent paid, and other small rents for pightles, tenements and meadows 'formerly paid to the tenant at Stiffkey Hall', which provide further evidence that subletting was common with these large tenancies.[131] Almost all Felton's letters to Townshend include some reference to the sale – or non-sale – of corn and hay, the prices to be realized, and the problems encountered.[132] For example, in 1680, 'the hay sold in the account within was for the yeare 1679, the corn for the yeare 1680. The hay for the yeare is yet to sell …' And then, 'That wch hinders the making of the yearly farm of Stiffkey £500 is in the low prices of corn, and the fall of the rent of the meadows, by reason of the abundance of clover and nonesuch yearly sowne'. The 'Money, Corne and Hay' remained 'in the custody of Richard Teasdale', placing Townshend at serious disadvantage. The correspondence reflects his unease. In 1682, requiring clarification, he wrote to Felton, 'I must ask Smith whether by agreement with Teasdale the rent was not come in cleer to me and he to allowe for charge I had bin at to plow and sowe the ground wch I presume I shall meet with in the Morston account … Nether do I find the corn Teasdale had of that ground brought to account yet.'[133] Townshend's distrust of the local bailiff, Samuel Smith, did not help matters.[134] In 1683, anxious for the future, Townshend enquired 'whether [Teasdale] be resolved to hier or not, if he will speak freely what he will do now at this very time, I will see to let it against his comes out, either by dividing it or in the whole, but think having houses I had better divide' . Teasdale was uncertain, 'whether he could find stock fit for the farm'. The farmall of 1683 shows that the sowing to halves agreement was still in force. In 1684 he paid a rent of £389.60, but by 1685, Stiffkey Hall had been taken in hand, with Teasdale still owing £350.[135] From this correspondence we can see that several options were considered: subdivision of the farm, which was eventually achieved in 1698, direct farming, or continuing with the existing arrangement of sowing the arable to halves. In fact, another letting to halves agreement survives from 1686 with Mr Johnson agreeing to sow the arable to halves, with Townshend paying half the cost of clover and nonsuch seed. This indicates that the practice remained an option, but there is no reference to it in the accounts. We can assume it was never implemented.[136]

So, why did Townshend not repeat the experiment? What was wrong with Teasdale's agreement? The first problem, as Townshend recognized, was the lack of clarity. Teasdale's rent for the pasture was set against his costs of cultivating the arable, but as corn remained unsold Townshend lost both rent and sales, while Teasdale received payment for his labour. In technical terms, like Windham at Reepham Farm, Townshend lost the initiative to the tenant. Coupled with this, the accounting procedure at Stiffkey entrusted far too much to Teasdale and gave him every opportunity to cheat and evade. Secondly, the falling price of corn placed acute pressure on these fragile arrangements, raising the tensions between the parties. As at Felbrigg, chasing and checking up on Teasdale – not to mention Samuel Smith – was a distasteful and demeaning exercise for a man of Townshend's status. But, probably, the deciding factor was Felton's death at the end of 1682; the arrangement had relied heavily on his monitoring skills and accounting expertise. Just as important, Townshend trusted him, unlike his successors, Bartholomew Snelling and Samuel Smith. From that time, Townshend preferred, like his brother-in-law, to take vacant farms in hand and work them with bailiffs and direct labour. In 1685 he devised new management instructions to deal with the change of policy. Stiffkey Hall Farm did not make more money, it still caused trouble, but at least he regained the initiative.

The situation at Stiffkey Hall was resolved by establishing a dairy of 20 cows.[137] As at Intwood Hall, the dairyman leased the herd and undertook to cultivate the arable with the rent he paid for the cows set against his farming costs. In 1685, Townshend allowed Nick Cooke £55 to buy cows for which he paid £45 a year rent for the cows, a sum including pasture, fodder and buildings, and £9 for two extra closes.[138] Cooke's wages and bills amounted to some £150 a year, so the £55 rent for the cows provided Townshend with a useful cushion against the falling price of corn. Like Brewster's agreement at Intwood, it did not involve any sharing of the corn crop. Leasing dairies, and paying the tenant to farm the arable, became a recognized strategy, continued by the trustees, for dealing with vacant farms. At Langham, further inland, a similar strategy was followed when Henry Parson's large holding failed in 1689. In this case, 303 acres of low value arable was leased to R. King for £91.50 and

169 acres, worth £77.07 was taken in hand and equipped with a dairy. John Raby rented 13 cows for £26, and was paid £28 a year, which included wages for a man to work the arable. The number of cows was important. When the herd was reduced to 11 and then six cows, the costs of farming increased and the benefits, in terms of profits and manure for the farm, declined. Neither Townshend nor the trustees bothered to adjust the number of cows and the rates paid for cultivating the arable, as Brewster did at Intwood to ensure that the enterprises at least broke even. Nevertheless both arrangements survived until 1698, when Charles Townshend, as part of his new strategy, leased out Stiffkey Hall and Langham Farm.

Apart from the break up of Henry Parson's farm, the situation at Langham and Morston remained relatively stable. A notable feature of this bailiwick was the predominance of peasant-sized holdings. Even the large holding shown in 1670 is misleading as it was leased to three tenants in partnership, while Parson's farm, a recent creation, proved unsuccessful and was swiftly reversed.[139] It is clear from the leases granted by Townshend in 1682, and renewed by the trustees in 1689/90, that no further attempt was made to consolidate holdings.[140] The explanation for this policy lay in a recognition and acceptance of local conditions. As at Hunstanton, where smallholders enjoyed by-employments in textiles, these men and women on the north Norfolk coast were not wholly dependent on agriculture. They could scratch a living in a variety of ways, including fishing, boat building, seafaring and carting goods, not to mention grazing sheep and cattle on the marshes, and dabbling in butter and cheese production.[141] They enjoyed certain rights and were particularly assertive: both Felton and Thomas Ward complained of the non-payment of rent, the constant demand for further abatements, and the need to appoint 'a trusty person' to ensure that 'the tenants manure their lands according to their agreement'.[142] It made sense, as far as possible, to work with the grain of this resilient peasant community, dividing large holdings, granting abatements, encouraging partnerships between tenants, and tacitly accepting subletting and informal agreements. Thus, in 1679 Felton advised Townshend not to treat with a tenant at Langham whose partner had died, until 'we see if Youngman can get an able partner'.[143] Faced with the sharp fall in corn prices in 1689, the trustees continued this policy. New leases supported tenants by offering substantial abatements in the first few years, which often amounted to more than half the rent, with incremental increases thereafter. For example, George Mattsall paid £9.45 in the first year, £22.45 for the next four years, rising to £24.45 for the final two years. In effect, the strategy was not dissimilar to the practice at Mack's Farm, Dilham where Windham accepted half the product for the first two years. Rather than meddle with a difficult tenantry, the preference at Langham was for a straightforward subsidy. But this does not exclude the possibility that farming to halves occurred between tenants.

The problems at East and West Rudham (Table 5.7) were long standing and typical of those properties situated on the brecks to the northwest of Raynham, which included the smaller estates at South Creake and Stanhoe.[144] As we have seen, in this area of very light soils, tenants struggled as corn prices fell from the mid-1660s. With reduced incomes, they were unable to stock the foldcourses on which fertility depended. At Stanhoe, Edward Money would only agree to a rent of £100 if Townshend bought his sheep for £120 and leased them back to him.[145] The Townshends' strategy in the Rudhams was dominated by the need to manage effectively four foldcourses: Coxford, Manyards, Grannohill and Great Grounds. In 1662 each foldcourse was leased with a large holding, the idea being to integrate them into a more intensive farming regime, enclosing the brecks, and ending the extensive practice of outfield and infield farming. However, as tenants failed the Townshends were forced to revert to a policy of intervention and subdivision.[146] Thus in 1665 Grannohill foldcourse was leased to a partnership and later divided into two tenancies; Great Ground brecks were split into three holdings and let sown with wintercorn, oats or summerley at 10s per acre, while at Coxford Abbey, Townshend shared the cost of cultivation with payments to William Clarke for 'carrying out half the corn crop' and 'one half of the muck'.[147] Finally, in 1673, Manyards was divided between two tenants. When these tenants failed the foldcourses were taken in hand and restocked. Problems were not confined to the largest holdings. The 1667 account records no less than ten holdings vacant, out of a total of 46, a few in their '3rd year without a tenant'.

Leases from the 1670s indicate the scale of the problem and the reforms undertaken. Farms were re-ordered with a balance of meadow, pasture, brecks, infield and outfield lands to support a mixed farming regime. Townshend provided new buildings, sheep for 'tathing the said brecks', and stipulated a precise course of husbandry. Felton drafted these bespoke leases to meet the needs of individual holdings.[148] For example in 1672, Harper and Bennet, to compensate for four brecks lying in Manyards foldcourse, were allowed to break up 30 acres of summerley, 'specified by the bailiff', but not to take more than three corn crops together. In 1676 Robert Barrett was required to occupy his infield lands in six shifts, 'to sowe in course and so to lie for grassland',

Table 5.7 The changing structure of holdings at East and West Rudham, 1662–1730

Farm rents	Number of holdings										
	1662	1667	1682	1683	1684	1693	1701	1704	1709	1720	1730
<£1	8	6(4)	5	6(7)	6	10	8	9	7	5	6
£1 – £5	15	9(2)	9	4(3)	3	7	4	6	4	3	1
£5 – £10	10	12(2)	6	5	3	2(3)	3	3	1		
£10 – £20	6	8(1)	3	4	2	1(2)	1	1			
£20 – £30	2	1	1	2	2	2	1	1			
£30 – £40			1		2	4(1)	1	1	2	1	
£40 – £50			1	1	1	(1)	1	1		1	
£50 – £100	2	1	3	1	2	1	2	1			1
£100 – £150	1	1	2	1	1	1(1)	1	1			
£150 – £200	2	1(1)		2	2	(1)	1	2	3	2	
>£200							1	1	1	2	4
Holdings let	46	39	31	26	24	28	24	27	18	14	12
In hand and let to halves		(10)		(10)		(9)					
Total rent (£)	858.40	628.00	841.65	722.70	829.35	436.25	796.00	825.10	833.90	963.15	1257.15
In hand and let to halves (£)		196.30		119.00		377.35					

Source: Townshend MSS, RAD/C3–4; RAD/B2–3; RAS/A1/6.

one of the six shifts was to be summertilled. He was not to sow oats on the infield lands or more than four crops without giving the same a summertill . The leases contained the usual penalties against ploughing up pasture and sowing out of course and to observe the requirement to spend all muck, compost, straw and stover on the premises. The strategy worked. By degrees, Barrett took on parts of Coxford Abbey at a lower rent; in 1677 William Ramsey, a small local landowner, accepted the lease of Manyards with its foldcourse, and J. Denny and J. Daniels continued to share Grannohill.[149] But the Great Ground proved more difficult. A lengthy correspondence between Townshend and Felton explains why.[150]

The deterioration of the Great Ground foldcourse at West Rudham stemmed from the failure to find a tenant to supervise the management of the brecks. When Mr King surrendered his lease in 1665, the Great Ground breck was divided between three men, but without supervision the cultivation was neglected and it became 'over run with bracks and furs', hindering the grazing of sheep and causing them to get 'staggers' and die. 'When it was with corn it had sheeped well and proved one of the best wether flocks in the country'.[151] Under-stocking accelerated the vicious downward spiral. With advice from Peter Stringer, Felton devised a management plan, to identify the cycle of decline and propose a solution, 'Now for the Remedy'.[152] At his own cost, Townshend needed to clear, plough and sow the brecks and take in more closes – the brecks were 'so bad no man will hire it'. The best way to improve the rent was to lease the foldcourse with a farmhouse, barns, sufficient infield lands and the tithes of West Rudham. The barns would hold all the corn, 'muck will arise from the straw of the corn of both the brecks and the tithes and will muck nere 100 acres of infield land wch may be taken from tenants that [neglect] yo lands to mend their owne – 50 acres whereof might be laid to it. And thus ordered it would be a good farm in time, and would improve both the foldcourse and the tithes and soone answer all the charge'.[153] Anticipating that initially no tenant would be forthcoming, Felton recommended 'yo Lordship must hire a man and put him into Pope's house, purposefully to plow the brecks and such lands as shall be laid to it, wch ly convenient, and yo. Lordship cannot lose by so doing, and ... will by this means bring the Foldcourse in request and make it as good a farm as any'.[154] Felton did not advocate letting to halves, as he did for the improvement of the Raynham meadows, but given the tradition of halving crops and holdings at the Rudhams it is likely that something of that nature occurred.

The scheme appeared to work. In 1682 John Blade accepted a seven year lease for Pope's messuage and the Great Ground breck, paying in the first year, £112.66; the second, £130.66 and £138.76 every year thereafter.[155] Townshend agreed to provide 700 sheep to tathe 40 acres for corn. But Blade struggled from the outset, debts mounted and he left in 1696 owing £455. Other attempts at reform also proved short-lived. The farmall shows that in 1683, 55 acres, the lands late Bennet's, were divided into ten holdings and let to sow for a year, peas at 4s 6d per acre, barley at 5s and grasslands at 1s 6d. In 1684, Mr. Ramsey, the tenant of Manyards, and Denny and Daniels, who shared Grannohill, surrendered their farms. This failure prompted a further reappraisal. From 1684, Samuel Smith's accounts indicate a new strategy of hiving off the foldcourses and farming them in hand to ensure their proper management and adequate stocking rates.[156] Smith purchased sheep from the outgoing tenants, acquired more at local fairs, and brought in ewes and lambs from the Stiffkey flocks. By 1685 six foldcourses were managed as a fully integrated enterprise: the two Stiffkey flocks, the Morston foldcourse, and three of the Rudham foldcourses, while the farm land was either let in pieces or cultivated using direct labour. In 1686 Townshend's directions to his new steward, Thomas Ward made particular reference to his farms in hand and flocks at Stiffkey and the Rudhams, confirming that a coherent policy had been settled for the management of these difficult holdings.

Thomas Ward's appointment in 1686 brought a new clarity and direction to estate management at Raynham, immediately obvious from his meticulous accounting procedures. For example in 1687 he provided Townshend with an abstract summarizing the performance of the estate, which included a list of the several farms in hand.[157] In compliance with Townshend's wishes he accounted for them separately, and provided a bundle of accounts for the farms in hand and foldcourses alongside those for the six bailiwicks. The 1687 account for the Rudham farms in hand shows clearly the high costs of direct farming, the fragile nature of the returns and the risks involved. At the head of the account, Ward listed the acreage cultivated (406 acres of brecks and field land with the tithes and barn) and the rental value of £165.95. The profits received in money, £93.91, included sales of corn, grazing and stock, while profits not paid in money, £148.72, covered the wintering of beasts, feeding of sheep on turnips, delivery of bullocks, corn, hay and straw to Raynham and allowance for seed.[158] The total profits, £242.67, fell short of the costs of cultivation and disbursements at £276.96 In fact the loss was far greater. Ward noted £183.05 owing to him on the account, £276.96

less £93.91, indicating the generous calculation of the payments made in kind. Returns worsened sharply during the collapse in prices of 1689/1690, prompting Ward and the trustees to adopt a system of letting to halves at the Rudhams, South Creake and Toftrees. Whether Horatio Townshend would have pursued the same route is a matter of conjecture, but Thomas Ward would most likely have convinced him, as Felton did at Stiffkey in 1677.

Responsibility for the Raynham Estate passed to trustees in November 1687, when Horatio, 1st Viscount Townshend died leaving a 13 year old heir. Of the four trustees, Colonel Robert Walpole of Houghton played the leading role; it was to him that Edward Le Strange sent the accounts for his perusal, and he who monitored the working of the letting to halves agreements.[159] Walpole had long influenced affairs at Raynham, particularly in the Rudhams, where Townshend lands lay intermingled with his own, often acting as arbitrator and adviser in disputes and valuations. It was in the accounts for East and West Rudham that letting to halves first appeared in 1690. Given the examples from Houghton, it is possible that Walpole influenced its introduction with Ward modifying his accounting procedures for the farms in hand to accommodate the practice. The accounts from 1690 to 1693 show that the new system took time to bed down (see Appendix 2). The bundle of papers surviving for 1691 refers to 'Rudham farmes late in hand, now let to J. Butler and W. Roads', but in the actual account, Ward charges himself 'with the moyeity of the Corne', cross referenced to an agreement drawn up in August 1690. In the following year, the bundle title refers to 'Rudham farmes in hand and let to halfes'. In 1693, Butler and Roads appeared before Walpole for not meeting the terms of the agreement – this is the only instance of a reprimand – evidence of Walpole's close surveillance of the operation of the agreements, and the effectiveness of his leadership.[160] By this time the scheme was also established at South Creake and Toftrees. The proportion of leased farms, farms in hand and those let to halves in 1693 can be seen in Table 5.8. In the Rudhams, the total value of lands let to halves amounted to £200 a year, including the rent of the foldcourses. At no stage were the letting to halves agreements amended or extended. Farms came in hand, such as Coxford Abbey with its foldcourse, warren, tithes, farmland and pightles, and were simply broken up and re-let at reduced rents. Letting to halves was part of a limited strategy, designed to improve the brecks and keep holdings in cultivation until tenants could be found who were willing to shoulder the risk of these large, difficult farms.

Table 5.8   Farms leased, in hand and let to halves on the Raynham Estate, 1693 (£)

Bailiwick	Farms leased	Farms in hand	Land let to halves	Total rental value
The Raynhams	1,147.75	142.00		1,289.75
Toftrees	181.45		143.55	325.00
East & West Rudham	436.25		377.35	813.60
Stiffkey	83.50	389.25		472.75
Langham & Morston	374.78	77.05		451.83
South Creake	64.85		259.25	324.15
Stibbard & Ryburgh	75.00	45.00		120.00
Totals	2,363.58	653.30	780.15	3,797.03

Sources: Townshend MSS, Estate Account 1693, RAS/A2/1–8.

The extension of letting to halves to South Creake and Toftrees is an indication of the severity of the situation in 1689/90. In 1662 the estate at South Creake had been leased almost entirely as a single holding to a substantial tenant, and consolidation continued until 1680 when Edmund Greene rented the farm for £276 18s 2d.[161] This success, compared to the Rudhams, can be attributed partly to its more favoured position, but also, it would appear, to a history of good management. Besides the two foldcourses and brecks, the farm included a balance of pasture, meadow and arable closes with adequate buildings and stock. Nevertheless, by 1686 Greene was in difficulties, 'his failing was that he was weakly horsed so that some of the land was not tilled'. Once again lack of working capital was the problem. Ward recommended his replacement by Simon Moss, ' a ritch fellow', who promised 'honestly' to pay his rent and equip the farm, 'tenants of such rent are not plentiful in these times and not always free'.[162] This proved to be the case when Moss died, leaving his widow as tenant. By 1689 she and her new husband owed £332. The trustees did not break up the farm, preferring to support the couple allowing them £110 for summerley, tillage, straw, chaife and colder. In the following year, however, they took the entire holding in hand. In 1692, they restocked the two foldcourses and managed them directly, while John and Richard Fox agreed to 'let to halfes' the principal portion of the farm as Table 5.9 indicates. The remainder they leased to six smallholders.

Toftrees, north-east and adjacent to the Raynham estate, was a recent acquisition and the changing structure of its holdings is shown in Table 5.10.[163] The terms agreed with the vendor, William Ruding, were generous, with Townshend agreeing to lease the estate back to him for £330 a year, and to bear the cost of extensive repairs and new building,

Table 5.9   The changing structure of holdings at South Creake, 1662–1730

Farm rents	Number of holdings						
	1662	*1668*	*1684*	*1693*	*1701*	*1720*	*1730*
<£1	1						
£1 – £5	2	2	2	3	1	1	1
£5 – £10	1	1		4			
£10 – £20	2	2	2				
£20 – £50	1			3			
£50 – £200							
£200 – £250	1						
£250 – £275		1		(1)			
£275 – £300			1				
£300 – £400					1		
>£500*						1	1
Holdings let	9	6	5	10	2	1	1
Let to halves				(1)			
Total rent (£)	313.65	309.85	315.17	64.85	308.90	518.90	532.10
Let to halves (£)				259.25			

*includes the purchase of Captain Dove's estate in 1707, worth £190 a year
Source: Townshend MSS, Farmalls, RAS/A1/6; Estate Accounts, 1670–1730, RAD/B2–B3; RAS/A2/1–8; RAS/A3/1–8; RAD/C3–C5.

Table 5.10   The changing structure of holdings at Toftrees, 1684–1730

Rent	Number of holdings						
	1684	*1687*	*1693*	*1701*	*1712*	*1720*	*1730*
<£1			2		2	3	2
£1 – £5	2	2	2		3	5	4
£5 – £10					1		
£10 – £30	1	1	1(1)		3	1	2
£30 – £40						2	2
£40 – £50	1		1				
£50 – £60			1				
£60 – £100			1		1		
£100 – £300			(1)		2	2	3
>£300	1	1		1			
Holdings let	5	4	8	1	12	13	13
Holdings in hand and let to halves			(2)				
Total rent (£)	395.00	365.00	181.45	335.00	560.28	604.65	614.15
In hand and let to halves (£)			143.55				

Source: Townshend MSS, Farmalls, RAS/A1/6; Estate Accounts, 1670–1730, RAS/A2/1–8; RAS/A3/1–8; RAD/C3–C5.

including a washhouse, brewhouse and dairy. He also lent him £400 against his first year's rent. Despite this injection of much-needed capital, Ruding struggled to pay, and collect, his rents. As a small landowner, he had leased small farms to his own tenants and this arrangement continued, effectively introducing a layer of sub-tenants.[164] This structure, embracing subletting, becomes more apparent as Ruding acquiesced in the break up of his large tenancy in 1691. Table 5.11 shows ten holdings leased directly to the sub-tenants, including a dairy and two holdings 'let to halfes' to Thomas Beaumont and Smyth. Ruding retained a limited area around the house sufficient to graze some stock, but his newly built dairy, with about 25 cows, the trustees leased to Mr. Pigg for £58. The two good farms and the handful of tenements, pightles and shops were let at fixed rents. Thus, sharefarming existed alongside fixed rent tenancies. The portions let to halves included the part of Hempton closes, not retained by Ruding, and Beaumont's holding. These intricate arrangements at Toftrees – it is the only property which combined the leasing of a dairy with letting arable to halves – indicate that Ruding played a leading role in shaping them, indeed the trustees may have inherited existing agreements. Ruding's involvement in this limited way continued until 1698, when Charles Townshend leased 'the whole of the town of Tofts' back to him at the old rent of £335. The example at Toftrees

Table 5.11   Extract from Thomas Ward's account for Toftrees, 1691

Mr Ruding gent. for pte of Hempton Closes ye house gardens & Grays meadow being part of his former farme	17	5	00
R. Pigg for the Darey farme being pte of the former farme late in Rudings use	58	00	00
E. Gooding for a messuage & 32 acs of infield lands & 30acs of break late in Rudings use	60	00	00
R. Hipkin for a tenement and shop	00	14	00
H. Grey for a pitle	00	10	00
The foldcourse pte of Hempton Closes, the farme late Sam Smyth pte whereof is lett to halfes soe I am to account for them by themselves	00	00	00
Widow Harman for a tenement & pitle	3	0	0
Rich Smyth late Rudings and Cosens use now lett at	40	00	00
E. Foaker for divers lands	2	0	0
Tho. Beaumont the £12 p.an is lett to him to halfes to be acct for by itself ...	00	00	00

Source: Townshend MSS, RAS/A2/1–8 for 1691 account.

shows once again how formal leases to large tenants often concealed evidence of sharefarming with sub-tenants; it also demonstrates how success depended heavily on the expertise of trusted individuals like William Ruding.

From Thomas Ward's general accounts, we can begin to assess the performance of the foldcourses, farms in hand and those let to halves from 1687 to 1697.[165] In Table 5.12 the first column provides an indication of rental value, with the following columns showing the return on each holding, for eight years, from 1687 to 1697. Losses were most common when the foldcourses were taken in hand and restocked. The profits remain uneven as Ward managed the foldcourses as an integrated enterprise, moving stock from one to another. For example, sheep were drawn from Stiffkey to stock the upland flocks at South Creake, the Rudhams and elsewhere, inflating the returns from those foldcourses. The priority was not the sale of lambs and wool, but to ensure adequate tathing (manuring) for the arable lands. Nevertheless, the profits from the foldcourses increased from £317 in 1692 to £745 in 1697, and the number of sheep rose to over 12,000, back to the level of the expansive 1630s. The most noticeable improvement was the performance of the Rudham farms, coupled with the profits from the Rudham foldcourses. This two-pronged strategy of taking the foldcourses in hand and farming the arable to halves clearly worked.

In contrast to the steady progress of the farms let to halves, the returns from Stiffkey Hall Farm and Langham Farm appear erratic. At Langham this can be explained by variations in the numbers of cows leased, and in the receipts from the sales of corn and stock. At Stiffkey the dairy proved more reliable, but problems with the sale of corn could severely dent profits. In 1688 corn worth £200 remained unsold. Unlike letting to halves, where these risks were shared, with direct farming the estate shouldered all the costs and losses. A noticeable difference between the two strategies is the fat bundle of vouchers accompanying the farms in hand, compared to the slim bundles for those let to halves, demonstrating quite clearly the higher management costs involved with direct farming. Nevertheless, these farms made respectable profits; at least they did not suffer the huge losses encountered at Felbrigg. From time to time, farms at the Raynhams came in hand: 'Arnold's at South Raynham'; 'Dutchman's', the cow grass farm at West Raynham; and two farms in East Raynham. These holdings were easily turned round by leasing dairies and using direct labour, until a large tenant accepted a lease. There was a ready demand for these

Table 5.12 Abstracts of Thomas Ward's accounts of farms in hand and let to halves, 1687–97, showing profit and loss of enterprises

Foldcourses and Farms	Rental	In hand		In hand, let, and let to halves						
		1687	1689	1691	1692	1693	1694	1695	1696	1697
Manyards West Rudham	72.00	48.85	47.85	28.38	43.08	61.00	104.35	103.90	90.26	127.10
Grannohill West Rudham	65.00	38.90	39.35	18.00	38.70	56.10	81.2	98.05	78.26	84.35
Rudham Great Ground	54.00	–26.41	59.40	41.32	28.06	87.10	131.95	86.97	56.91	82.40
Coxford East Rudham	with farm							–208.60	–24.16	10.25
Morston Foldcourse	70.00	97.95	157.90	70.00#						
Stiffkey Hogg Flock	36.00	21.80	31.55	07.20	29.01	12.50	8.90	30.00	56.56	–7.50
Stiffkey Ewe Flock	45.00	00.15	18.90	12.47	22.28	07.90	22.10	07.00	14.17	–9.16
Stibbard Foldcourse	45.00		–160.28	18.40	25.28	26.65	15.12	37.65	45.60	92.57
Shereford Foldcourse	32.00		30.90	15.30		22.75	40.36	78.42	81.46	94.57
Creake East Flock	with farm			–9.86	27.55	22.75	52.27	21.10	21.20	29.35
Creake West Flock	with farm			–19.80	59.97	77.10	104.76	80.85	135.45	158.31
Tofts Foldcourse	with farm			–19.46	43.57	48.13	45.90	98.15	81.45	83.20
Rudham Farms	406.00	–183.50	–56.10	35.62*	58.45	96.91	221.76	116.16	88.11	49.00
Rudham, Blade's Farm	77.52								75.15	69.36#
Stiffkey Hall Farm	389.00	194.23	33.78	30.32	276.37	270.70	285.07	85.87	197.40	228.47
Langham Farm	77.10			27.76	31.57	94.37	57.92	16.50	17.07	32.87
Creake Farm	259.00				49.71*	100.06	117.81	98.41	117.16	88.96
Toftrees Farm	143.55				35.55*	68.14	67.26	56.95	68.19	59.75
W. Raynham, Dutchman's	40.00				21.27	–11.50	47.33	40.00#		
E. Raynham, Tubbing's	70.00				50.62	73.70	60.00#			
E. Raynham, Elmer's	65.00				63.57	65.00#				
Scarndale, Horningtoft	22.00						11.73			

#Let   *Let to halves
Source: Townshend MSS, RAD/C3; RAS/A2/1–8.

increasingly well equipped and attractive holdings. None was taken in hand after 1693.[166]

Having examined the detail, how successful was farming to halves at Raynham? Success should be judged against the earlier and later performance of these holdings: in the 1670s and 1680s no tenant with sufficient capital could be found for Stiffkey or South Creake; the Rudham farms failed repeatedly; and at Toftrees William Ruding was forced to relinquish his tenancy and sell his stock to repay his debts. As at Felbrigg, it was a strategy adopted *in extremis*, when all other approaches had failed. Yet, the table summarizing the performance of holdings shows immediate and positive results, notably at the Rudhams where the run of losses was turned into sustained profitability. The restocking and direct management of the foldcourses played a major role, but letting to halves, by sharing the risk, provided tenants with the vital support and an incentive to maximize their efforts. From a landowner's point of view, it reduced the costs of management, as the Stiffkey and Langham accounts clearly demonstrate, and the losses were shared with the tenant.

The scheme continued without modification for seven years until Charles Townshend assumed control of the estate. He promptly terminated the agreements, first at the Rudhams and Toftrees in 1697, followed by South Creake in 1698, and sold the flocks to the incoming tenants. This may reflect a fundamental change of attitude or even disapproval of the system. On the other hand, it may indicate that seven years of letting to halves was sufficient to ease tenants into large holdings. In 1697 John Butler accepted a five year lease for his lands in West Rudham and Manyards foldcourse paying a rent of £235, and £16.50 a year interest for his flock, while Robert Tinckler entered Grannohill foldcourse and William Road's lands at East Rudham, paying a rent of £155 and £13.50 for his flock. The following year, William Tinckler took a lease of Coxford Abbey with the foldcourse for £183, and William Burrage added the Great Ground to his lease of Blade's Farm.[167] These farms continued to progress: by 1720 the rent for the Rudhams had increased from £818 to £963, an 18 per cent increase since 1697, and the four main tenants had repaid the principal sums for their flocks.

At Toftrees and South Creake the system of letting to halves proved equally successful. As we have seen, in 1698 William Ruding renewed his lease for the 'whole town of Toftrees' paying £335, while Phillip Tubbing agreed to pay £300 for the great farm at South Creake. When Ruding's estate was broken up on his death in 1712, a substantial increase in rent was secured over the next seven years, from an average of 5s 6d to 10s per acre; overall the rental rose from £335 to £604, nearly 80 per cent since 1697. The increase at South Creake was less dramatic, as it remained in Tubbing's tenure. These outcomes suggest that letting to halves played a significant role in effecting improvement at Raynham in the late seventeenth century. The consolidation of holdings attempted by Horatio Townshend in the 1670s and 1680s had proved premature; tenants were unable to bear the burden of paying rents and stocking large farms as prices continued to fall. Tenancies were broken up, foldcourses divided and various stratagems devised to farm these holdings. In the 1690s, the trustees instigated a system of letting to halves which, by supporting and sharing the risk with tenants, enabled them to build up capital and stock. In this way, it laid the foundation for the large tenancies secured by Charles Townshend after 1697. The fact that he achieved the transition so swiftly and successfully is testimony to the effectiveness of letting to halves.

Why was letting to halves more successful at Raynham than at Felbrigg? A notable feature at Raynham was the lack of tension, false starts and financial disasters. This can safely be attributed to Thomas Ward's exemplary management and accounting procedures and the leadership provided by the trustees, notably Robert Walpole. Objectives were clearly defined and carefully controlled. Letting to halves was never more than a contract to grow specified acreages of corn, very different from William Windham's imaginative, risky schemes. In no circumstances did Ward and Walpole link letting to halves with dairy farming. At Toftrees the dairy lease remained entirely separate from Thomas Beaumont's agreement for the arable. Unlike at Felbrigg, the initiative was never surrendered to the tenant. The experience at Raynham is proof that when the practice was well monitored, letting to halves was an extremely cost effective method of securing improvement, as Felton had explained to his master in 1679. Inadvertently, he also highlighted the principal weakness of letting to halves. Success depended on management skills of the highest order. Not every landowner was fortunate enough to employ a Timothy Felton or a Thomas Ward, or had useful neighbours like Robert Walpole of Houghton, to administer affairs during a long minority.

Finally, how significant was farming to halves on these four Norfolk estates? Given the incomplete and erratic nature of the evidence this is a difficult question to answer. With the exception of Blickling, where alternative devices were adopted, farming to halves was used as an estate policy on leading west Norfolk estates throughout the seventeenth century, and at Felbrigg at least from the 1670s to the 1690s. The estate records only document the relationship between landlord and tenant, but we can reasonably assume that farming to halves occurred on all these estates between tenant and sub-tenant. Chance references show that it was used at this level at Dilham, on the Felbrigg estate, in the 1650s and at Langley, on the Blickling estate in 1718.[168] At Toftrees, on the Raynham estate, the failure of the tenant in 1691 revealed sub-tenants leasing a dairy and farming portions of arable to halves. At Hunstanton, on the other hand, evidence showed two tenants paying a fixed rent to Sir Hamon Le Strange in the 1630s, and at the same time, entering into a sharefarming agreement with his son for cultivating the new drained marshland. Thus, as in parts of Europe, fixed rents co-existed with sharefarming, sometimes on the same holding, operating at two distinct social levels, with large tenants occasionally paying fixed rents and subletting portions on sharecropping terms. The case studies confirm that behind the formal fixed rent tenancies, which appear in the estate records, lay a hidden structure of subletting and sharefarming.

As in Europe, farming to halves seems to have been particularly associated with the seventeenth century. The price rise of the late sixteenth century forced landowners to commercialize the management of their estates. They needed capital to invest in new ventures, which entailed expenditure on labour, livestock, infrastructure and equipment. They also needed tenants to farm the newly improved holdings. These tenants, lacking capital themselves, required support in the initial stages of improvement. As we saw at Hunstanton and Raynham, in the first half of the century, in some situations landowners resorted to farming to halves, sharing the risks of draining marshland and intensifying grain cultivation on the brecks. The alternative was to concede low fixed rents in return for the tenant shouldering the entire risk. Not only was this course less profitable in the short term, but landowners lost control of fragile improvements and jeopardized their investment. This became a particular problem when prices fell from the mid-1660s. As we saw at Raynham, large tenants entered a downward spiral: unable to stock their flocks, grain yields declined and their income collapsed, forcing them to surrender their tenancies. Similar outcomes were evident on the Blickling and Felbrigg estates, particularly on the light soil farms, which had been the subject of earlier improvement.

The owners tackled the problem of low corn prices and vacant farms in different ways. On at least three estates, they financed the diversification into dairy farming, establishing and leasing dairies to incoming tenants. When it came to farming the arable, they adopted slightly different strategies: sowing to halves, offsetting the costs of cultivation against the rent for the dairy, paying contract rates for the job, or simply providing a guaranteed price for the corn. At Blickling and Felbrigg, intervention of this nature invariably preceded tenants accepting fixed rent tenancies, albeit at reduced rents, but at Raynham, and possibly on other west Norfolk estates, the problems were deeper and more fundamental. The trustees at Raynham tackled the issue by taking all the flocks in hand, and letting substantial portions of the worst affected farms to halves. What had been an *ad hoc* arrangement, used to overcome a temporary difficulty as at Stiffkey Old Hall in the 1680s, became a systematic approach to solving the problem of light soiled farms in the 1690s. The accounts show that the strategy proved successful, re-establishing the fertility and profitability of these difficult farms, reversing the downward trend which had affected them since the 1660s. At the end of the century, Charles Townshend was able to lease the farms at fixed rents, without granting substantial concessions to the tenant.

The case studies show that in the transition towards a capitalist agriculture, farming to halves offered a way of easing tenants into new holdings, sharing the cost of improvement and building up the capital of landowner and tenants. Above all, it mitigated the risks of farming in a period of economic turbulence and political instability. While its significance on these estates is difficult to calculate, it clearly performed an important role at Hunstanton in the early part of the century, at Felbrigg in the 1670s and 1680s, while at Raynham, where the evidence is more consistent, it was a key feature throughout the seventeenth century, helping to support a sometimes fragile farming regime. The evidence from Raynham shows graphically that farming to halves was adopted at two social levels, by the Townshends as a last resort, but more especially between tenant and sub-tenant as a cost-effective way of further exploiting the holding, sharing the risk of large farms and supporting the fixed rent payable to the Townshends. The question is why, if farming to halves could prove so effective, did landowners abandon the practice so abruptly in the early eighteenth century, when the economic climate would appear to have dictated otherwise?

6
Sharefarming Disappears from the Documents in the Eighteenth Century

What happened to letting to halves, and other forms of sharefarming, in the eighteenth century? In the 1770s Adam Smith, as we have seen, dismissed *métayage* as a practice so long in disuse in England that he knew no English name for it, while Arthur Young, a decade later, claimed that the very success of English agriculture rested on the absence of such a system.[1] How had a practice, prevalent less than a century before, disappeared from view? None of the county reports, drawn up in the 1790s for the newly formed Board of Agriculture, contain any reference to *métayage*, and as far as we know, Smith and Young are the only English authors to have commented on the practice at this time.[2] They were not challenged in their observations until John Stuart Mill wrote favourably on *métayage* in the 1840s, criticizing Young for his 'extremely narrow view of the subject'.[3] However, Mill accepted Young's contention that *métayage* was not an English practice, and on those grounds he did not advocate its adoption in England, preferring a revival of peasant proprietorship. Significantly, when proposing peasant proprietorship, Mill felt obliged to explain the term and its virtues as Englishmen were 'profoundly ignorant' of the idea. Steeped in the language of agricultural improvement, led by landlords and tenants, they had progressed so far beyond their peasant origins, that they had no knowledge or understanding of the social condition of peasants or mode of life, and little interest either. For shaping this outlook, we need look no further than Arthur Young and his contemporaries, who tended either to ignore or denounce any remnant of peasant farming, their avowed purpose being to create a new consensus based on the benefits of large farms, enclosure, and the capital intensive landlord-tenant system. So, our task in the eighteenth century is doubly difficult. First, with Norfolk landowners disengaging from farming to halves at the end of the seventeenth century, we lose a valuable source of material. And secondly, with the propagandizing nature of agricultural writing in the late eighteenth century, attention is concentrated on the virtues of the landlord-tenant system to the exclusion of other tenurial relationships.

Despite these difficulties, we know that forms of sharefarming, including farming to halves, continued into the eighteenth century. This chapter will show how the large tenancies introduced on large Norfolk estates in the late seventeenth and early eighteenth centuries concealed agreements between tenants and their sub-tenants. Correspondence, bills of sale, family settlements and legal documents continue to provide glimpses of a hidden layer of activity with large tenants, effectively acting as middlemen, leasing dairies to cowkeepers and parcels of land to smallholders, often at halves, at least until the 1720s. Cornish probate inventories show that half shares continued into the 1730s. Evidence of cow leasing persists in farm diaries, county surveys and later in the reports of parliamentary commissions. Using material from Dorset and other parts of the west country, the practice will be examined to show how it underpinned the development of the dairy industry in that region. Finally, these findings will be assessed and compared with the latest literature on land tenure in England and on sharecropping in Europe.

In the last chapter we saw how Norfolk landowners having resorted to forms of sharefarming in the seventeenth century, finally abandoned the practice in the late 1690s. By 1700 evidence of letting to halves and other informal agreements has disappeared from estate records, although we know it continued between tenants and sub-tenants. The decision to withdraw from active involvement in farming was part of a process whereby landowners distanced themselves from their tenants, in favour of a uniform system of fixed rent tenancies administered by professional estate stewards. The explanation for this trend has many strands. At an economic level, over the preceding decades of depressed prices, valuable lessons had been learnt. Negotiating with tenants, controlling budgets and enforcing new strategies in such an adverse economic climate required professional skills of the highest order. It was not always in the landowner's interest to intervene in the farming activities of their tenants. As tempers frayed and working capital was lost, it became evident to landowners that creating and equipping large farms, likely to attract enterprising men with capital and farming ability, was a far safer option. When prices failed to maintain the modest recovery achieved in the mid-1690s, landowners accepted lower rents to secure men of substance willing to shoulder the risk of farming. With the poor weather and bad harvests of the 1690s, followed by repeated slumps in prices in the 1700s, landowners became increasingly disillusioned with agriculture and eager to withdraw.[4]

The new Land Tax may also have influenced landowners' decision-making. It is worth remembering that when it was introduced in 1689 it was the first regular tax on landed incomes and coincided with a sharp fall in corn prices.[5] Levied initially at one shilling in the pound, rates quickly rose to four shillings, which reduced any incentive to maintain rents at artificial levels. Imposed on net profits, it made sense to simplify management, let rents drift downwards and support tenants by investing in fixed capital. Increasingly, Norfolk landowners pursued this policy, delegating estate administration to full time professionals and the business of farming to well capitalized tenants. As we have seen, the process was accompanied by radical changes in the composition of estate records. Landowners required summaries of income and expenditure, and lists of tenants and rents, not explanations of routine business. In this streamlined format, details of farming activity and informal agreements disappeared from the documents, just as they did when estates were leased in the fifteenth century. Furthermore, the landlords' responsibility for the payment of the Land Tax meant that the farming activities of tenants did not attract the attention of tax officials, unlike in France where peasants paid the *taille*, so instances of subletting or informal agreements between tenants went unrecorded and remained hidden from view.

This process of withdrawal, which we have identified in estate management, has also been associated with a much wider cultural shift in the attitudes of the gentry.[6] The catalyst was the political settlement of 1688, which freed landowners from the concerns that had preoccupied their forbears since the 1630s. By the 1690s they had achieved their objectives and could turn their attention to foreign policy, business and leisure. From that time, they created a distinctive gentry culture with its own rhythms and calendar, with extended absences in London and Bath, visiting friends and relatives. In his work on the Townshends of Raynham, Rosenheim drew attention to the withdrawal of the county elite from county government. While the Civil War generation conscientiously attended local business and served on the bench, the next generation delegated local affairs to their nominees, while they pursued a cosmopolitan lifestyle elsewhere. Robert Britiffe, political agent to Sir Robert Walpole, forced to appoint men of lower status, commented on the 'great fault of Mr Ashe Windham and other gentlemen'.[7]

The cultural perspective offers an insight into why estate management changed so abruptly in the 1690s, why the Civil War generation collaborated with tenants, and why their successors disengaged from such activities. Neither Ashe Windham, nor Sir Henry Hobart nor Charles, 2nd Viscount Townshend, nor Sir Nicholas Le Strange shared the anxieties of their fathers or grandfathers. Estate management no longer had to be subordinated to the broader interests of securing peace and harmony, it could be organized along narrowly commercial lines to serve the long-term interest of the new ruling and leisured class. Their fathers, marked by their experience in the Civil Wars, and influenced by the spirit of renewal evident in the 1660s, had turned their attention to domestic matters and the management of their estates. Declining prices, coupled with a government policy designed to promote agricultural improvement, provided landowners with every inducement to reform their estates. However, the supportive element of many initiatives, such as providing loans and sharing the risk of farming, suggest they were motivated by more than economic advantage. The cost implication of some of these policies, as prices continued to decline, led to their modification and phasing out from the late 1680s. What was different in the 1690s was the explicit desire of their sons to distance themselves from estate management and local concerns. Understanding these shifts in outlook is vital to our understanding of agricultural change, as it offers a corrective to the view that change was an automatic response to the movement of prices and costs.[8] The experience of Norfolk estates bears this out. The attitude of landowners in the 1690s and 1700s was very different from the 1670s and 1680s.

At a more practical level, the political settlement of 1688 had greatly advanced the position of landowners. As the new ruling elite, with control over the legislature, they acquired the power to pass laws to protect and promote their own interests. Bearing in mind the fines and confiscations suffered by landowners during the Civil Wars this was a great step forward, securing for them absolute private property rights and raising, very significantly, the capital value of their estates. Coupled with the development of the modern laws of mortgage, landowners were most advantageously placed in the land market. The attraction of land as a capital investment was further enhanced by an unreformed electoral system, where the right to vote was based on the ownership of land. As Parliamentary elections became a regular part of political life, landowners aspired to control Parliamentary seats, influencing government and building up patronage networks around their country estates. This is precisely what occurred on the great estates sin Norfolk in the early eighteenth century. As Offer explains, from the 1700s land acquired a high socio-cultural value, a positional premium, well in excess of the economic returns from agriculture.[9] This development had two important effects. First, it put land purchase beyond the reach of those who earned their living by farming, and secondly, it encouraged landowners to invest heavily and reduce rents to attract good tenants. The trend was reinforced by the need for landowners to secure an assured income to finance an increasingly lavish lifestyle. This, according to Stead, made landowners increasingly risk averse and eager to pass the hazards of farming on to their tenants.[10] In this way, the new order created the optimal conditions for both landlords and tenants. Landowners had no need of strategies like sharecropping, while tenants, benefiting from low rents, fixed capital improvements and new credit facilities, could generate their own working capital and profit from large farms. At the same time, young scions were able to pursue new interests and opportunities, with young Charles Townshend and his neighbour Robert Walpole leading the way. The English landlord-tenant system, with its well-defined responsibility for fixed and working capital, and rents fixed for terms of years, clearly suited the interests of both parties. For the way it encouraged investment, reorganization of farms and radical improvement, the system won the approval of the political economists and agricultural writers of the day. These changes, it has been claimed, amounted to a landlords' revolution.[11]

Yet, the landlords' revolution was not so neat as the foregoing analysis implies. As we have seen, large leases granted on Norfolk estates in the late seventeenth century and early eighteenth century frequently concealed agreements between tenants and sub-tenants. The evidence becomes more difficult to detect in estate records, yet, references, *en passant*, still occur in legal documents and correspondence. In the sale documents of the Langley estate, a scribble in the margin identifies lands farmed to halves, although the estate accounts simply recorded Mr Berney paying an annual rent of £474.[12] The particulars of the Blickling and Gunton estates, drawn up by Robert Britiffe for the marriage of his two daughters, in 1717 and 1732, acknowledged the existence of under-tenants.[13] But by far the most extensive evidence of the continuation of subletting and forms of sharefarming in the early eighteenth century is found in the Raynham archive.

The most explicit example concerns the Shipdham estate, purchased by Charles Townshend from Sir John Castleton in 1704. The estate, situated a few miles from East Dereham, was leased to a single tenant, John Stagg for £365 a year. The sale particulars recommended that the holding, 'to improve the rent', should be 'layd out in small farms'.[14] Although Stagg kept his farm 'in very good order', grew 50 acres of turnips and possessed stock worth the large sum of £1,000, he proved unable to pay his rent. Finally, in 1706, after many unfulfilled promises, he agreed to the sale of his stock as security for his debts.[15] The valuation shows that he owned three large dairy herds, including 43 cows and six milk trays 'at the farme late Fullers in Shipdham'; 42 cows and six lead milk trays 'in Rawlinson's use', and 45 cows and seven lead milk trays at the Park Farm in Shipdham, besides seven heifers, one bull, and 24 horses 'upon the farm of John Stagg'. But for the forced sale we would have no idea about the subletting and dairy farming. Stagg was a gentleman farmer, a trusted figure, and so long as he paid the rent it was of little concern to Townshend how he arranged his business. Simplicity of management was the objective, and as far as possible such tenants were to be encouraged and supported. As it happens, Stagg surrendered his tenancy in 1717. The steward's account for the following year shows the estate divided into seven holdings, reflecting the concealed structure of 1704, and the recommendation of the vendor that the estate should be divided into small farms. The rent increased from £400 to £507 following subdivision, indicating the size of the concession the Townshends had previously made to the single tenant in return for undertaking the risk of a large farm. It was, in fact, more profitable for the Townshends to break up the estate into smaller tenancies and lease them directly to several tenants.

The rentals, leases and memoranda, which survive for the period 1697–1701, show Charles Townshend's commitment to a policy of leasing large farms to substantial tenants at fixed rents.[16] From 1698, Ward was obliged to send his master a single page 'Abstract' summarizing the performance of the different bailiwicks, identifying non-paying tenants, and providing totals of profit and loss.[17] By 1701, Townshend had successfully leased the entire estate to a string of tenants, principally by offering rent reductions and investment in fixed capital. The rental in Table 6.1 gives the impression of a model early eighteenth-century estate, with large farms leased at fixed rents. But appearances are deceptive; behind the large tenancies a much more varied structure survived.

As we saw in Chapter 5, Charles Townshend reversed many of the policies adopted by his trustees. For example, at Stiffkey Hall and the principal farm at Langham and Morston, where the trustees had leased dairies to tenants and paid them to cultivate the arable, the farms in 1698 were leased to tenants at fixed rents. However, although the dairies disappeared from the accounts, the receipts for that year show that dairying continued. On both properties, the incoming tenants and several other men purchased cows, dairy, butter and cheese-making equipment,

Table 6.1   Comparison of rents and holdings on the Raynham Estate, 1693–1730

	1693		1701		1730	
Bailiwick	*Farms*	*Rent (£)*	*Farms*	*Rent (£)*	*Farms*	*Rent (£)*
The Raynhams etc	52	1,147.75	47	1,324.65	51	2,087.87
In hand	3	142.00		1		300.00
Toftrees	8	181.45	1	335.00	13	614.15
Let to halves	2	143.55				
E. & W. Rudham	28	436.25	24	796.00	12	1,257.15
Let to halves	9	377.35				
Stiffkey	10	83.50	16	423.55	8	424.60
In hand	1	389.25				
Langham & Morston	27	374.78	27	440.00	16	610.48
In hand	1	77.05				
South Creake	10	64.85	2	303.25	2	532.10
Let to halves	1	259.25				
Stibbard & Ryburgh	11	75.00	9	103.98	2	186.50
In hand	1	45.00				
Shipdham					7	507.70
Farms leased		2,363.58		3,725.45		6,220.39
In hand		653.30				300.00
Let to Halves		780.15				
Total Rental		3,797.03		3,725.45		6,520.39

Source: Townshend MSS, RAS/A1/6.

while four new tenants rented simply a meadow. The smallholders kept a few cows while the larger tenants leased cows to cowkeepers who specialized in butter and cheese making. As at Shipdham, the large tenancies concealed cow leasing. Like his father, Charles Townshend also encouraged partnerships. When he leased Stiffkey Hall in 1698, not only did he sub-divide the farm, but leased each half of the farm to a partnership, who also rented the Stiffkey flocks. He underwrote them, lending seed, granting loans for the purchase of the flocks, selling equipment, and providing horses. The Townshends, like the Le Stranges at Hunstanton, continued to support collaboration between tenants, leasing flocks and foldcourses to syndicates and partnerships, and breaking up large farms into more manageable tenancies, as a way of mitigating risk.[18]

The lease of the Toftrees Estate to William Ruding as a single holding for £335 a year, looked as if it concealed sub-tenants, and so it proved. A memo written by Ward, (unfortunately the heading and date is torn off), showed that 'Mr Ruding let off £139 to eight sub-tenants, and retained in his own use £196'.[19] In 1712 when Ruding died, the estate reverted to this fragmented structure. Four new farms were created, with the tenants receiving substantial concessions in the form of heavily reduced rents for the first three years. This was not letting to halves, but a direct subsidy to establish their farming operations. As at Shipdham, rents rose significantly, from 5s and 7s to 10s per acre, between 1712 and 1730, following subdivision into family sized farms.[20]

Further evidence on the Raynham estate reinforces the picture of a layer of sub-tenants underpinning large holdings.[21] For example, at West Raynham, when Curson's and Stringer's lands came in hand in 1707, smallholders and cowkeepers quickly came to the fore, leasing closes, cow grasses and a dairy of twenty cows for £45. In 1712 Thomas Sexton accepted a lease for the whole lot paying £110 a year rent, and from that time the cow grasses disappeared from the accounts. But when he failed in 1719 the holding was sub-divided between seven tenants, raising the possibility that subletting had continued between 1712 and 1719. Ward's correspondence also indicates the existence of sub-tenants at South Creake. When Philip Tubbing's lease came up for renewal in 1704, they actively considered subdivision as an alternative, but it remained let as a single holding. From 1707 Tubbing also leased Captain Dove's estate in North Creake, creating a farm worth over £500 a year. In effect Tubbing continued to act as middleman on this estate, as his family had done for decades. From these examples, we detect a transitory stage between sharefarming and full fixed rent tenancy, when landowners supported new tenants with rent remissions, loans and investment in fixed capital to get them started in a period when prices remained flat, while old tenants remained on concessionary rents, subletting dairies with the acquiescence of the landlord. Contracts remained flexible with the emphasis shifting from one to another depending on circumstances.

The situation at Raynham, and the other Norfolk estates, can be compared to developments in Europe at this time. Santos describes how sharecropping co-existed within fixed rent tenancies in southern Portugal from the seventeenth to nineteenth centuries.[22] The Alentego area, on which his study was based, shared many characteristics with parts of west Norfolk in the early eighteenth century: dry, light sandy soils, cultivated with sheep and corn, the land held in huge estates and divided into large farms leased to substantial tenants on fixed rents for terms of years. These tenants employed armies of labourers, and sublet portions to sharecroppers. Many of the farms were let to middlemen, so the details of sharecropping remain hidden, but Santos, using the records of a charity foundation, found evidence of the practice. He observed two types of contract: leasing for a term of years for a fixed contractual rent, and share-cropping for one annual cereal crop. Sharecropping, as at Raynham, was restricted to cereals with the sharecropper paying a percentage, usually a quarter, of the crop. Unlike the practice in England, the contract was sanctioned by medieval law, 'lords of vineyards and farms let them to be cultivated for half, third and fourth ... follows the nature and quality of the contract of sharecropping'.[23]

For the directors of the charity, leasing on fixed rents was preferable to sharecropping as it generated more income and guaranteed greater stability from the capital owning tenants. Sharecropping involved high surveillance and enforcement costs for landlords with large scattered estates. Only as a last resort would landlords choose sharecropping. For example, when a tenant had terminated his lease or a new tenant could not be found, landlords might sharecrop for a season in order to avoid a total loss of income, keep the farm in good heart and entice a tenant for the following years. This was a strategy familiar to William Windham of Felbrigg. When demand was weak, sharecropping arrange-ments could run for several years, but were replaced as soon as a suit-able tenant paying a fixed rent could be found. The farmer's interest in subletting to sharecroppers was similar to that of the landlord. Faced with high labour costs, sharecropping involved lower vigilance and enforcement costs and allowed the tenant to obtain an income on poor quality land with practically no investment. The quarters paid by the sharecropper amounted to a significant income, plus the value of clear-ing undergrowth. In this way, sharecropping formed an important aspect of the *latifundium* economy of Iberia. For the landlord, as long as the tenant kept the farm in good shape and paid his rent, subletting posed no problem and required no interference. This policy bears a striking resemblance to the strategy pursued by the Townshends at Raynham.

In his analysis of land tenancy markets, Santos showed that, during this period, two distinct and superimposed rent markets existed for the same land. First, the market for multi-annual leases for fixed rents, where the demand came from tenants and urban groups looking for a productive asset. Secondly, a more precarious sharecropping market existed, with rents varying in relation to the harvest, which involved plots of land for growing cereals, often land that required clearing after a long fallow, or in need of cultivation to fertilize the ground. The demand came from small peasant farmers who owned some working capital but not sufficient for the whole farm. Thus the lease market only realized part of the rent. It was common for tenants to sublet to sharecroppers in the secondary market part of the property rights they had acquired in the primary markets. In this way, fixed rent leasing and sharecropping co-existed functionally within a single system of production, just as it did at Raynham, and affected the different participants in different ways. For the large landlord, sharecropping was used as a contract of last resort, either for short-term contingencies or when effective demand on leases dropped below the offer, particularly in hazardous times, like the crisis and war-ridden seventeenth century. In such risky times fewer farmers had the resources to enter into fixed rent tenancies. The landlords had in many cases to share fully the risks through sharecropping contracts if they were to get tenants to keep the farms productive. For the tenant paying a fixed rent, subletting to sharecroppers earned a rent from plots he did not want to cultivate at virtually no cost. For the sharecropper, the contract provided affordable access to land for him to deploy working capital and family labour, within the comparative safety of keeping a share of the crop. It allowed him to maintain his status as a peasant farmer, accumulate capital to achieve upward mobility, and possibly to become a tenant in his own right.

This European perspective begins to explain how early modern farmers secured sufficient capital to take on large tenancies. Mingay identified this as one of the principal obstacles to the creation of large farms.[24] We saw at Raynham that holdings were frequently modified and subdivided to accommodate smaller tenants. Mingay wondered how prospective tenants raised the several hundred pounds required to stock and equip large farms. He suggested several possible routes: marriage to a wealthy bride, inheritance, and mortgaging freehold land and credit from local merchants; nevertheless, it remained something of a mystery. We would argue that for the accumulation of capital, forms of sharefarming should be considered. As we shall see, devices like cow leasing continued to perform the same role as sharecropping in the Alentego, providing additional income for large farmers, and a pathway for labourers to advance themselves by deploying family labour and accumulating enough capital to lease a tenancy. If the large farmer was a tenant, as was increasingly likely in the eighteenth century, there were also two rent markets existing on the same land, for precisely the same reasons, and with similar beneficial effects for all parties, as in the Alentego.

Not long after Arthur Young completed his *Travels through France*, condemning *métayage* and noting its absence in England, he was appointed secretary to the newly created Board of Agriculture. The formation of this body was prompted by the steep rise in corn prices in 1793 and the outbreak of war with France which threatened supplies. Led by Sir John Sinclair, its purpose was to galvanize the landed interest and promote improvement in farming. The first county reports appeared in 1794, often the result of hurried surveys by men not always fully acquainted with their region.[25] Despite these limitations, John Claridge in his *General View of the Agriculture of Dorset* provided a detailed account of 'The management of the dairy, as everywhere practiced in Dorsetshire'.[26] What follows is an almost identical description of the leasing of dairy cows to dairymen as we saw at Raynham, Blickling and Felbrigg in the seventeenth century. In 1815, in a second survey, William Stevenson elaborated on the system, including the comments of farmers, presumably in response to the first report.[27]

The Dorset system, as revealed by Claridge and Stevenson, only concerned the leasing of dairies. The arrangement was between farmer and sub-tenant, like that of Mr Stagg and his dairymen at Shipdham, on the Raynham Estate, in the 1700s. The dairies were leased for a year with the farmer providing the cows, feed and fodder, and letting the cows to the dairyman at a fixed price, agreed at Candlemas. The rent in 1794 averaged £6 a cow, depending on the region and the age of the stock.[28] The farmer found a house for the dairyman, allowed him to keep pigs, poultry and a mare to carry his butter to market. He retained the right to give the dairyman six months' notice. These terms remind us of Edmund Britiffe's agreements at Felbrigg, without the provision of sowing the arable to halves. In addition, the dairymen leased summer grazing or 'cow-leazes', the Dorset equivalent of cow grasses or pasture, for 30s per acre, at a stocking rate of between one to two acres a cow, the same as at Houghton in the 1650s and 1660s. As at Raynham, the cows entered the cow-leaze between the 5 April and 12 May, and stayed until about the 20th August, when they were allowed after-grass and then wintered on straw from 23 November (Martintide). They were dried off by the end of the year, and calved in the spring. Farmers generally let their cows in several lots of 20 to 40 cows; one farmer 'let 50 cows to a dairyman who ploughs and does other work at his leisure'. This passing reference may be a hidden example of sowing to halves. Stevenson noted that it was 'almost impossible to ascertain with any degree of precision the average produce of the cows, as the immediate interest of those people obviously forbids them to furnish correct accounts'. In other words, as always in these matters, they kept their affairs secret and hidden from prying eyes. A farmer from Corfe Castle highlighted the benefits of the system, although the 'fashion of letting cows to dairymen is not approved in other counties, but as the cows are milked and the butter and cheese made and sold by the dairyman and his wife and family, it is rational to suppose they will take more care of their own property, than they would of their masters, and thus it proves beneficial to both parties'.[29] In his final comments, Stevenson questioned the wisdom of large farms, where farmers leased as many as 100 cows to three or four dairymen, 'it would be much better for the country, and not much worse for themselves, if the size of their farms was diminished'.

Evidence of the Dorset system is not confined to the county surveys. In James Warne's farm diary of 1758, he noted that in March of that year, his father agreed to 'let Bovington dairy to old Haines, Mr Bewnel's dairy-man now at Coomb, 60 cows for £3 a cow'.[30] A young man at the time, he and his father, Joseph, a prosperous tenant farmer, occupied several farms, often moving to better themselves, leasing substantial dairies to cowkeepers. Horn provides a fuller description of this type of farming, showing how cow leasing existed from the early eighteenth century to the most recent times.[31] In 1714 Humphrey Weld of Lulworth Castle leased a herd of 30 cows and one bull for a rent of 45s a cow, comparable to the rents in Norfolk, while in 1754 the Rector of Symondsbury let his cows for £3 5s, similar to the figure paid by the Warnes. By 1794 rents had doubled to £6, and with wartime inflation, they doubled again to an average of £12 by 1815. In 1855, J.C. Morton, estimated in his *Cyclopedia of Agriculture*, average rents to be £9 10s, while Joseph Darby, writing in 1872, noted a range of between £10 to £12 a cow.[32] Specific examples included Walter Ross of Ibberton leasing 32 cows and heifers to brothers, John and Henry Watts, for an average rent of £12 10s in 1874, and John Butler of Tarrant Monkton letting his herd of 20 milkers for the same rent in 1882. Subsequently, rents declined in the 1890s, but recovered during the Great War to £20 a cow, a level that was maintained in the 1920s and 1930s. Although the practice became rare after the Second World War, an example was found of a dairyman paying £72 a beast in the 1970s. The movement of rent indicates that in the eighteenth century cow leasing was extremely profitable for dairymen, but in the nineteenth century, following wartime inflation in the price of cows, profitability rapidly declined making it a less attractive proposition for dairymen.[33]

The dairy agreements explain how the system worked.[34] The cow rent was paid on a quarterly basis, with the dairyman being allowed remissions for late calvings, normally expected to occur in early May. On his part, the dairyman promised not to milk any cow for longer than forty weeks from the date of calving; in the later part of the century, he often provided a quantity of cotton cake or additional fodder out of his own pocket. For example, in 1887, for the management of 114 cows, the dairyman agreed they were 'to be calved down in good condition, all the best to have corn and chaff at once with 3 tons of best cotton cake after calving'. The survival of the system can be explained by the benefits it offered to both parties. For the farmer, it relieved him of the responsibility of managing the herd, particularly important in a county dominated by sheep and corn farmers who rather despised cow keeping. Secondly, it gave him an assured income and was more profitable than using direct labour, as cowkeepers worked on very narrow margins. Thirdly, and linked to this point, cowkeepers used family labour, usually including a wife and daughters skilled in the arts of butter and cheese making – skills which could not be replicated by employing direct labour. From the cowkeepers' point of view, renting cows also had many attractions. First and foremost, it allowed a man with limited capital to set up on his own account. All he needed was his own dairy utensils, which were relatively inexpensive and enough cash to pay his first quarter's rent. Paying a fixed rent for the cows gave him every incentive to maximize his efforts and produce a high quality product. In 1895, the Royal Commission on Agriculture commended the system for the opportunities it offered to labourers eager to progress up the farming ladder from the position of a stockman on wages, to dairyman, and thence to tenant farmer. Indeed, it provided almost the only opportunity for labourers to better themselves.[35] Census returns confirm the point, showing cowkeepers moving from dairy to dairy every three or four years, taking on larger herds.[36]

There were, however, disadvantages to the system. The dairyman, paying a fixed rent for the cows, wanted to make as much profit as possible, so he might not pay full attention to the long-term interest of the cows, the pastures, the fixed equipment and the buildings. Cowkeepers were noted for their mobility, moving from herd to herd, extracting the maximum profit. This point was made by William Marshall who condemned it as 'injurious to an estate: as tending to let down the buildings and the fences of farms thus occupied by under tenants: who have not so permanent an interest in keeping them up as a lessee'.[37] But the alternative to cow leasing, was not a lessee, but a waged labourer who would have even less interest in the welfare of the cows, or in producing a quality product and maintaining the infrastructure of the farm. Cow leasing suited the task and the times. The survival of the Dorset system was linked to the production of quality butter and cheese, employing skilled family labour at low cost. The practice started to decline with the coming of the railways, when Dorsetshire dairying turned to the production of liquid milk for the London market. The simpler task made it much more profitable, and easier, for farmers to manage their large herds with carefully directed and well remunerated wage labour. At the same time, the increased costs of equipping dairies, and from the Second World War the rising cost of fodder, made cow leasing an increasingly risky business for the small man with limited capital. In many ways it was better, and certainly safer, for him to sell his valuable skills as a dairyman for good wages and 'perks', which often included a cottage, garden, free milk and meat. Comfortable and secure, a dairyman on good wages, stayed in his employment, to the benefit of the cows, his employer and himself.

Cow leasing was not confined to Dorset; it extended to most parts of the west country, including the cheese and butter making areas of Hampshire, Wiltshire, east Somerset, east Devon and Cornwall, where it underpinned the development of a commercial dairy industry in that area from the late seventeenth century.[38] Favoured by soil and climate, the trend to the west country may explain why Norfolk landowners moved out of dairying in the early eighteenth century, and why similar decisions were made in Suffolk.[39] In the Wiltshire Vale, dairying became increasingly specialized in the seventeenth century, with yeomen buying up smallholdings and hiring 'sharemilkers' to milk the cows.[40] The county report for Wiltshire of 1794 confirms that cow leasing existed on the Downs, 'in the corn part of the district', where 'great farmers frequently let their dairy cows by the year', to use the water-meadows and permanent pasture.[41]

Another example from the county reports shows a variant of the Dorset system in Cornwall, linking us back to the evidence of half shares and moieties in the probate inventories. Worgan described farmers letting 'cows to labourers and poor people at £6 or £8, for 7 or 8 months – 4, 6, 8 or 10 cows to each person'. The arrangement worked as follows: 'The hirer pays his cow rent by milk and butter, for which he finds a ready market in this populous district. When a cow approaches her time of calving, the farmer is obliged to take her, and provide the person with another, flush in the milk. These cow renters generally have a piece of ground allotted to them by the farmer on which they grow potatoes – with these and scalded milk, which has yielded cream for butter, they fatten a great many porkers.'[42] This example shows cow leasing performing a slightly different function from the commercial role it played in Dorset and Wiltshire. In the Cornish economy, dairying was not important, but in parts of west Cornwall on the high grounds, large portions of waste allowed farmers to overwinter and rear stock. The breeding cows were brought down from the commons, back to fields full of grass in early spring for calving. After weaning, the farmer retained the calf and let the cows. Apart from 'an oat sheaf' to give them a boost at calving time, the cows survived without additional fodder. This simple arrangement provided the farmer with an outlet for his cows in milk, some extra income for labourers, and supplemented the diet of Cornish fishermen and miners.

Finally, how far did the Dorset system of cow leasing conform to sharefarming? With the dairyman paid a fixed rent for his cows rather than sharing the produce, cow leasing is not sharefarming in the strict, technical sense of the word. However, sharing milk produced on a daily basis is not easy and even in France *métayage* agreements placed all kinds of restrictions over the sale of butter and cheese off the farm.[43] Sharemilking agreements made little headway before the development of dairy co-operatives in the 1890s. In the daily collections of milk from the farms, they recorded individual outputs, so they were able to divide the produce through the monthly milk cheque. In other words the dairy companies assumed the classic role of monitoring the product, traditionally considered to be the principal burden and risk of share-farming. In cow leasing, by providing working capital for the dairyman in the form of cows, and particularly in taking responsibility for herd replacements, the landowner or tenant farmer, whoever was leasing the cows to the dairyman, shared the risks of dairy farming in a very direct way. We remember William Windham's frustration with his dairies at Felbrigg, and difficulties in collecting the rent for the cows. Cow leasing was a form of risk sharing, if not sharefarming, finely tuned to the needs of the participants, the technology available and the circum-stances of the time. With the evidence from Norfolk and the west country, it is possible to claim that cow leasing, with its sharecropping elements, facilitated the commercialization of dairying in England from the late seventeenth century.

From the evidence so far, we can see that farming to halves, and other forms of sharefarming, continued into the eighteenth century and beyond. This experience conformed to European practice as identified by Santos in southern Portugal.[44] Far from being mutually exclusive, share-cropping and fixed rent tenancies were functionally complementary, with landlords and tenants drawing rent, at different levels, from the same land, and labourers gaining access to the land and capital. The parties continued to benefit from sharecropping in different ways. For landlords, it was used as a short-term contract when demand for fixed rents was weak. For tenants, subletting plots on sharecropping terms or leasing dairy herds was more profitable than hiring labour, and for dairymen it provided an opportunity to accumulate capital and the possibility of securing a tenancy. In this way, sharefarming continued to exist as a hidden institution, a flexible and informal device working within the more rigid system of fixed rent tenancies, often helping to underpin the large heavily capitalized farms.[45]

These findings do not, however, conform to recent work on English land tenure. In a style reminiscent of Arthur Young, Offer and Stead make the explicit assumption that sharecropping played no role in the English tenurial system.[46] Quoting Adam Smith, Stead abruptly dismisses its existence, while Offer associates it with the poverty-stricken areas of eastern and southern Europe; in his 'ladder of tenures', sharecroppers occupied the lowly position between wage labourers and tenants. In contrast to the experience in the rest of Europe, Offer's purpose is to explain why fixed rent tenancy became dominant in England, and how English landlords were able to maintain such a potentially precarious system for so long. Stead develops these points, and explores the ways English tenants mitigated the risks of farming which they bore under this system. At no stage does he consider the possibility that sharecropping might have formed part of their strategy. A puzzle for these historians is why well capitalized English tenants endured these risks and did not aspire to farm ownership, the top rung of the farming ladder, like their European counterparts, and their compatriots in Australia, New Zealand and North America, where the prospect of ownership had lured settlers to the colonies. In fact in England, as Offer points out, the movement was down and not up the ladder of tenures, as small owner-occupiers invariably sold their farms to raise the working capital necessary to lease the new large well equipped tenancies.

Ladders of tenures were deployed by the eighteenth-century political economists, Marxist historians, notably Goubert, and modern economic historians, including Offer, to explain the relationships between those that owned and farmed the land. Table 6.2 shows our version, with the rungs of the ladder ascending from wage labourer, sharecropper, tenant on fixed rents, to owner-occupier. Implicit in the ladder of tenures is the shifting of risk from the landowner to the tenant. With wage labour, the landowner accepted the risks of farming and the burden of management. This gave him flexibility and complete control, but monitoring labour could prove expensive and irksome. With sharecropping, the landowner shared the risks and farming costs. The prospect of keeping half the crop motivated the sharecropper and reduced levels of

Table 6.2 A ladder of tenures from the perspectives of landowners, farmers and labourers

Perspective of landowners	Ladder of tenures	Perspective of farmers and labourers
ROLE: To sell land to tenant	OWNER OCCUPIER	ROLE: Provides fixed and working capital, and labour. Takes all risk and responsibility for the farm
PROS: No burden; no cost gain		PROS: Whole crop; capital asset to pass on; complete control
CONS: No income; no asset		CONS: Takes all the risk; and needs to raise all the capital to invest on farm
ROLE: Provides fixed capital, limited management and financial support in difficult times	TENANT ON FIXED RENTS	ROLE: Provides labour and working capital; pays rent; takes responsibility for the farming
PROS: Certain income; control over capital; delegated management		PROS: Whole crop; landlord provides fixed capital; abatements in bad years
CONS: Lower returns		CONS: No capital asset to pass on
ROLE: Provides fixed capital; share of working capital; shares management and risks of farming	SHARE CROPPER	ROLE: Provides labour; share of working capital; responsible for farming
PROS: Reduced management and monitoring costs; incentives to labour; prospect of higher yields and returns		PROS: Incentive in half the crop; capacity to build up capital, stock and share risks of farming
CONS: Bother of collecting shares; uncertainty of income; issue of trust		CONS: Only half the crop; disincentive to invest; interference irksome
ROLE: Provides fixed and working capital; management of farm and supervision of labour	WAGE LABOURER	ROLE: Provides labour
PROS: Complete control; flexibility and maximum returns		PROS: No responsibility; flexibility; regular wage
CONS: Burden of management, high costs of motivating workforce		CONS: No incentives; little scope for advancement; no assets

management for the landowner. With fixed rent tenancy, the landowner received a fixed payment from the tenant which gave him a guaranteed income. In return, the tenant kept the entire crop, accepted the risks and gambled on making a surplus. With peasant ownership, all benefits and risks accrued to the occupier. The question for Offer is why wealthy tenants preferred paying fixed rents to ownership.

Offer's explanation lies in the institutional framework which so advantaged English landowners. As we saw, the political settlement of the 1690s had given them control over the legislature and guaranteed absolute property rights, which made land a secure and attractive investment. As values rose inexorably, small landowners and peasant farmers were forced out of the land market. In fact, rather than buying land, they were better off selling their holdings and using the proceeds to finance large tenancies, already enclosed and equipped with new buildings. They stood to benefit as landowners competed for tenants with ability and working capital, lowering rents, investing heavily and granting abatements when times were hard. Landowners effectively subsidized the activities of their tenants, and in this way, they shared the risk of farming, even if they did not divide the product. From this perspective, as Offer concedes, the difference between the landlord-tenant system and sharefarming is more blurred than we suppose. Tenants liked fixed rents, as it left them with the entire profit, and yet, in times of need, they were cushioned by leniency over rents and injections of landlord capital. Landowners also preferred a fixed rent as it gave them an assured income, and at the same time, they received the full benefit of fixed improvements in terms of the capital value of their estate.

Mingay, amongst others, draws attention to the scale of support landowners offered to tenants in the first half of the eighteenth century.[47] Low corn prices, disease in sheep and cattle, and general uncertainty forced landowners, like the Duke of Kingston, to tolerate arrears, grant abatements, reduce rents and offer all sorts of inducements to attract and keep tenants. He observes that concessions and expenditure on fixed improvements, closely reflected the price of corn, rather than interest rates. For example, before 1750 expenditure on buildings averaged 4 per cent of estate income, but after 1760, when prices rose, the figure fell to under 1 per cent. Expenditure remained low during the French wars, but increased significantly in the 1820s and 1830s, as prices fell. Similarly, responsibility for the payment of the Land Tax shifted away from landowners after 1760. This type of intervention shows landowners explicitly sharing the risks of farming with the tenants. The support for tenants was finely tuned and applied in a precise and flexible way; very different, from the blunt instrument of sharecropping. From this perspective, the landlord-tenant system can be viewed as a form of risk sharing with outcomes similar to sharefarming. The downside was that the system was built on the assumption that English agriculture would always provide adequate returns to both landlord and tenant.

From this analysis it becomes clear that so long as English landowners were prepared to subsidize their tenants, there would be little evidence of sharecropping in England. It existed in the seventeenth century before the landlord-tenant system was fully evolved and reappeared in the early twentieth century as the landlord-tenant system started to decline. Its disappearance in the eighteenth century coincided with a time when more and more land was held in large estates, and landlords leased their farms on fixed rent tenancies, sharing the risks with tenants through abatements, investment and forbearance. Stead argues that risk for tenants was principally mitigated by the availability of insurance, new forms of credit, improved transport systems, choice of crops, better livestock and mechanization.[48] This no doubt was true, but it does not explain the return of farming to halves in the early twentieth century. We would argue that the high level of landlord support, which provided a cheap and internalized form of credit, remained an important aspect of risk sharing and is the main reason why there is so little evidence of sharefarming in England from the 1720s to the 1900s. The chronology shows that risk sharing was embedded in different types of contract, irrespective of whether the contract was a fixed rent tenancy or farming to halves. It was not the case, as Stead argues, that tenants on fixed rents bore the entire risk of farming or that landowners only shared the risk when they resorted to explicit forms of sharefarming. The English landlord-tenant system was a risk sharing device suited to the times and the needs of both parties; when these changed, the parties reverted to other ways of spreading risk.

From a European perspective, Santos makes similar points and invites us to consider risk sharing as a continuous variable across contractual forms.[49] The risk sharing model of the tenancy ladder, put forward by Offer and others, whereby risk shifts from landowner to tenant, is misleading in that it implies that risk is a distinct variable that exists in sharecropping but not in fixed rent tenancy. Risk sharing in agrarian contracts went far beyond the frontier of sharecropping and well into the territory of contractually fixed leases. Santos shows how nominal fixed rents diverged from the effective rents paid, which are similar in pattern to the evidence in William Windham's Green Book shown in Table 5.1. Over ten-year periods the proportion of rent foregone in lease contracts correlated to the proportion of farms let on sharecropping contracts, supporting the idea that rent reduction and sharecropping were two related institutional devices with a general risk spreading practice. Both devices co-existed during the high risk periods, but with somewhat different timings. Until the 1640s deferring payments took the lead, with little resort to sharecropping until the 1630s as risks mounted and tenants were impoverished by wars. In the 1700s, when acquittals were not sufficient, sharecropping tenancies climbed again for a short time. In later crises, risk sharing was mostly contained by resorting to abatements. With some modification in the early eighteenth century, this pattern observed in the Alentego resembled closely the experience found on the Norfolk estates.

These observations led Santos to trace an evolutionary path from sharecropping in the late middle ages to effective fixed rent tenancy in the late eighteenth and early nineteenth centuries. For nominal fixed tenancies to emerge from a sharecropping environment, and for risk sharing within a nominal fixed rent tenancy to fade out, three things needed to happen: first, a reduction in agricultural risk, which included a change in crops, more efficient land allocation and technological improvements; secondly, the development of efficient credit and insurance markets; thirdly, the emergence of a distinct stratum of tenant, with sufficient wealth and credit status to be able to finance fixed rent tenancies. As Stead demonstrates, these factors were achieved in England in the eighteenth century, with the added impetus of landowners gaining political control which allowed them to shape the institutional framework in their favour. In the Alentejo, the evolution of agrarian contracts passed through several stages which can also be compared to developments in England. In the 1400s sharecropping abounded as farms were divided and leased as plots. With leases for lives tenants were expected to sublet plots on sharecropping contracts. By the end of the 1500s, farms were leased on fixed contractual rents, with sharecropping being used either as last resort by the landlord, or as a subletting arrangement by the tenant. During this period of growth and rising prices, tenants were prepared to take the risk of a fixed rent. The seventeenth century was a regressive phase, as prices tumbled and war ravaged the region. Nominal rents tumbled and sharecropping returned as a relatively frequent arrangement. Once these crises were over, fixed rent tenancies were re-established, with fewer acquittals and write-offs, until by the end of the eighteenth century, they were almost extinct. We recognize the similarities with developments in Norfolk. In the nineteenth century, the experience diverged as legal reforms in Portugal, in favour of tenants, facilitated full landownership, while poorer tenants, unable to make the transition, continued as sharecroppers.

In England, the situation was somewhat different. The vast wealth accumulated by English landowners, the dominance of large estates, and the degree of control they exerted through Parliament, enabled them to maintain their position and resist significant land reform, until the early twentieth century. The next chapter on the nineteenth century shows how they managed to keep control for so long, despite the pressures from below for land reform, which included elements of sharefarming.

7
Profit-sharing and Land Reform in the Nineteenth Century

In the early nineteenth century the favourable conditions for landowners and tenants which had existed since the 1760s ceased with the end of the Napoleonic Wars in 1815 and the resumption of cheap imports of corn from abroad. With falling prices, English landowners used their political power to secure protection through the Corn Laws legislation designed to keep prices artificially high. The debate over its imposition generated deep divisions within English society. For the radical press it was proof of the inequities and inefficiencies of the English landlord-tenant system; it was claimed to be an institution formed not so much to increase yields, but to reduce costs and make profits for the producers at the expense of labour.[1] The newly enclosed farms were not more productive per acre, but required fewer buildings, made optimum use of equipment and labour, and saved landowners the trouble of dealing with small under-capitalized tenants. From a landowner's point of view, this form of capitalist agriculture, so efficient in the use of capital and labour, made sense, but the cost to labour in terms of welfare could be severe.[2] Unlike the much derided peasant farms of France, the improved farms of England were unable to absorb the surplus rural population. When prices fell and English farmers cut their costs, landless labourers bore the brunt. Social unrest, culminating in the Swing Riots of 1830–1, required the government to seek remedies and others to offer solutions. These included the provision of allotments for labourers, a return to peasant proprietorship and profit-sharing.

This chapter begins by identifying the growing criticism of the landlord-tenant system and the demands for land reform from radicals and political economists. During this period, Robert Owen launched his ideas for industrial co-operatives while in Ireland, and an Owenite landowner, again in Ireland, experimented with profit-sharing in an agricultural context. Most significantly, J.S. Mill articulated the case for *métayage*, citing examples in Italy and France, but he did not advocate it for England, as 'the exigencies of society had not naturally given birth to it'.[3] This observation directed land reformers towards peasant proprietorship, rather than co-operatives and sharefarming. Modest recovery in the 1850s and 1860s gave landlords and tenants some relief, but the pressure returned as prices slumped in the 1880s and labourers drifted from the land. The second section will focus on attempts to stem the tide and show how a few enlightened landowners experimented with profit-sharing. Comparisons will be made with the revival of *métayage* in France and finally with the development of sharefarming in colonial New Zealand. The chapter will draw on the later surveys of the Board of Agriculture, the extensive literature on land reform, the reports of the nineteenth-century commissions on agriculture, the writings of philanthropic landowners and modern accounts from New Zealand.

The difficulty with this chapter is that no archival evidence of farming to halves in England has been found for the nineteenth century. The profit-sharing, introduced by philanthropic landowners, did not constitute sharefarming in the strict sense of the word, being a sharing of the profits, after landlord and labourer had extracted their costs in terms of rent, interest and wages, rather than a sharing of the produce (Type B in our taxonomy). Profit-sharing was essentially an incentive payment to labourers, rather than a contract to share the produce between landowner and farmer. Despite the absence of evidence, however, inferences can be drawn as to why farming to halves is likely to have survived in the nineteenth century, though it probably continued to decline. As we saw in the last chapter, the landlord-tenant system, based on the payment of fixed rents, developed in a way which suited both landowner and tenant. The landowner received a guaranteed return, while the tenant received generous support in terms of fixed capital investment and abatements of rent in times of difficulty. In reality, the English had devised a system which shared the risks of farming. In the eighteenth century, these risks were further mitigated by reductions in transactions costs, including improved marketing networks, the growth of country banks and the increased availability of credit and insurance. From these sources, tenants could borrow the working capital to stock and equip their farms, and were able to protect their business ventures. Improved crop rotations, selective breeding of livestock and more sophisticated farming techniques also helped to reduce the risks.[4] Progress on these fronts continued in the nineteenth century, providing a secure, familiar and institutionalized framework for landowners and farmers alike. In other words, there was less reason for resorting to schemes like farming to halves.

Despite these developments, farming to halves seems to have survived. There is written evidence of its existence amongst small owner-occupiers in the early twentieth century, in precisely the form it existed in the early modern period. Thus, we can reasonably assume that the practice continued unobserved, possibly in more limited ways, in the nineteenth century. Reed reminds us of the continuing significance of small farmers in rural communities, and the way historians have tended to overlook the activities of this social group.[5] Howkins makes a similar point; 'whole groups of fundamental importance to rural history of the 19th and early 20th century have been written out of history'.[6] This was particularly true of small family producers and worker peasant farmers, who were more common even in 1900 than standard accounts would suggest. As late as 1895, the Agricultural Returns show that 80 per cent of farms were under 100 acres.[7] This figure included men and women who combined farming with other occupations; typically, local trades people, craftsmen, smallholders, businessmen and the clergy. In a village, they were often numerically significant, even though they might not own or occupy the bulk of the acreage. When documentation improves, in the early twentieth century, we find these people often associated with farming to halves, making informal agreements with farmers to supply stock, and with labourers to farm their holdings.

In parts of Surrey, Kent and Sussex, where heath and commons still flourished, Reed also touches on these informal arrangements and the role they played in rural societies. He describes an economy of small-scale agriculturists gaining access to land through subletting, taking crops for a season, buying winter keep and paying in kind and labour for goods and services. He argues that such diverse activity, by sustaining low wages and low cost enterprises, underpinned the development of a capitalist rural economy. He does not mention farming to halves, but the impression is of a peasantry, much more akin to the French than we have been led to believe. Reed noted the observations of a French agriculturist:

> the agricultural prosperity of England is generally attributed to large farming, but ... ideas on the subject are much exaggerated... . No doubt there are very large farms, just as there are large estates; but these form by no means the majority. There is a multitude of farms under the middle size, which would pass for such even in France;

and the number of small tenants is infinitely greater than the number of small proprietors ...[8]

These remarks, by a foreigner, show quite clearly the distortions that existed in English agricultural writing in favour of large farms and great estates with the emphasis on the uniqueness of the English experience. English historians have continued these themes, and tended to ignore strands, such as peasant farming, which link us more closely to the European experience. An attempt will be made to remedy this omission in Chapter 8. In this chapter, of necessity, the focus will be on the types of sharefarming associated with landowners trying to motivate and ameliorate the conditions of their landless workforce. For reasons that will become clear, *métayage*, or farming to halves and produce sharing, was not on their agenda.

No less an advocate of improvement than Arthur Young drew attention to the adverse impact of enclosure and large farms on the condition of labourers in rural parishes.[9] While the benefits had been immense, in terms of increasing corn production and providing regular employment, labourers had suffered from the loss of common rights. In his county report for Norfolk, Young compared the situation with that of a decade earlier. At Fincham, 'As to cottage cow-keepers they are all over: many before enclosure, but the allotments all thrown to the farms, and in this respect they are much worse situated'.[10] He identified particular problems on the Fens where drainage and enclosure had wiped out the livelihoods of traditional communities. At Marshland, Sneeth and Fen, 'the poor people who turned cows, geese and ducks upon the common, without possessing rights, have suffered, as in so many cases, and it is to be regretted that some compensation is not in all cases provided by the Act ... the immense system of labour created is worth far more than such practices, still many individuals injured without any absolute necessity of being so'.[11] Young did, however, single out certain landowners for praise. Sir Thomas Beauchamp-Proctor at Langley had marked out lots of marshland for his 24 cottagers to keep cows; they paid rents to him of no more than £2 2s, for good gardens, small grass fields, and a 'place by the river assigned for mowing fodder for their cows ... Too much cannot be said in favour of such a system'.[12] Similarly, at Heacham, where the Le Stranges still owned land, houses owning common rights had been awarded two acres of middling and the right to stinted commons, '12–15 little and very comfortable proprietors and renters of small plots from 2–10 acres who have cows and some corn ... a remarkable instance, and I cordially wish it was universal'.[13] In both these villages we found examples of farming to halves in the seventeenth and early eighteenth centuries. The disturbances in East Anglia, which erupted in 1816, were concentrated mainly in the Fens, comprising riots and protests against enclosure and drainage schemes. Further outbreaks in 1822, against the use of threshing machines, were suppressed by the military with substantial loss of life.[14] These uprisings were the precursor to the more widespread Swing Riots, which finally forced the government to take notice of the problems of rural deprivation.[15]

The government responded by passing several pieces of legislation between 1819 and 1845. All these measures included some provision for allotments as a basis for improving the physical and moral condition of the rural poor: the Vestries Act of 1819 gave parishes the power to buy or hire land to provide employment; the Inclosure of Wastes Act, 1832, increased the provision of allotments by allowing parishes to enclose up to 50 acres of waste; the Fuel Allotments Act, 1832, compensated labourers for rights lost through enclosure, while the General Enclosure Act, 1845, required, as a condition of enclosure of waste and common, that allotments should be provided for the poor. These laws, according to Burchardt, were an act of reparation to assuage the deep sense of damage and the tangible sense of guilt amongst landowners for the deprivation suffered by labourers. Why then, given the sentiment of landowners, did this legislation prove to be of limited effectiveness? The main reason was the inappropriate agency entrusted with enforcement. Allotment provision was seen as an aspect of poor relief and the responsibility of parish officials, who had no incentive to add to their unpaid burden. Moreover, as most of them were farmers, they were not eager to surrender land, provide distractions for their labourers, or weaken their bargaining position by giving them a measure of self-sufficiency. In this head-to-head situation, between employer and employee, landowners were reluctant to intervene. As legislation remained permissive until 1887, it was easy for farmers to thwart local attempts to provide allotments; if they failed, they could simply refuse to employ allotment holders. Farmers were particularly opposed to large allotments on which labourers could lavish their time and energy; they preferred to grant 'potato grounds' on an annual basis, reaping the benefit of spade cultivation. The loss of status, and the blurring of social distinctions between farmer and labourer, was also an issue, threatening the carefully constructed hierarchy within the village. Schemes did not always win the support of labourers, who were often suspicious of landlords' intentions and resented the further distinctions made between deserving and undeserving poor.[16] Mill understood their misgivings. In his view allotments were a cynical way of making the poor grow their own poor rate.[17] In this way, the English landlord-tenant system had helped to create a social structure and a culture which made collaboration between the classes difficult. Nevertheless, as we saw in Dorset and Wiltshire, collaborative schemes, like cow leasing, did survive, and a popular demand for land and co-operative ventures continued to exist.[18]

The remedies supported by the ruling elite in the nineteenth century were not aimed at promoting sharefarming ventures between landowners and farmers. However, agriculture was influenced, if only at the margins, by the co-operative movement led by Robert Owen and the French idea of profit-sharing which had been advocated by Turgot in the 1770s, as a way of providing incentives for waged labour. Taylor drew attention to these early initiatives in his book *Profit-Sharing between Capital and Labour*.[19] Besides German and Danish examples, he described the scheme adopted by the Owenite Irish landowner, Mr. Vandeleur on his Ralahine Estate, County Clare, in the early 1830s. The context for this experiment was the rising rural discontent in Ireland in the 1820s, which culminated, as in England, in the agrarian crisis of 1830–1. Vandeleur, a friend and disciple of Robert Owen decided to engage his disaffected labourers in a scheme of 'participatory agriculture'. The scheme was managed by Edward Craig from Manchester, a committed Owenite.[20] The agreement involved the whole workforce of 52 men, women and children. Vandeleur provided the land, buildings, implements and stock, and paid the labourers daily wages at the ordinary rate; they farmed 622 acres, 326 acres of which were arable. The Association, made up of all the members, paid him £900 in rent for his land and interest on his capital. The net profits belonged to the Association, but 'were expended not in individual distribution, but in purchasing stock from Mr. Vandeleur and other worthwhile objects'.[21] The experiment proved a great success, but was abruptly curtailed, after just two years, by the sale of the estate to repay Vandeleur's gambling debts. Nevertheless, it provided an inspiration to the languishing co-operative movement, and paved the way for the Chartist Land Plan. As Taylor explained, the singular achievement of the scheme was the motivation of the workpeople, who developed a shared interest in the well-being of the estate and the community. But as a strategy of estate management, it remained an isolated experiment, firmly associated with Irish issues, extreme conditions and a most unusual landowner.

Land reform, throughout this period, remained a highly contentious issue as landowners defended their privileges against increasing opposition from all other sections of society, including, not only labourers, radicals and political economists, but manufacturers and the middle classes, forced to pay the price of protection. The crucial aspect of this crisis of the countryside was that the landed interest controlled the political response, and was strong enough to withstand challenges for fundamental land reform. Timely concessions, such as the encouragement of allotments, adjustments to the Corn Laws in the 1820s, and, indeed, the sacrifice of protection in 1846, were sufficient to fend off moderate critics. More advanced ideas, based on the redistribution of land and common ownership, made little headway. These schemes, associated with the eighteenth-century revolutionary thinkers, Thomas Paine and Thomas Spence, were presented as part of a radical political agenda intent on undermining the existing social order. Similarly, the co-operative schemes, pioneered by Owen, were written off as the work of romantics and extremists, a sentiment reinforced by the failure of the Chartist Land Plan in 1851. From our perspective, the interest of these schemes lies in the demand for land from the working classes and the desire for co-operative forms of agriculture.

The Chartist Land Plan has been largely dismissed as Utopian, nostalgic and escapist, but it is worth remembering that at its peak in 1847 it attracted 70,000 subscribers eager to rent the two acre plots offered by the newly formed National Co-operative Land Company.[22] Despite the ridicule heaped on it by contemporaries, there was nothing fundamentally wrong with the idea of working men combining together to purchase land, forming a company and leasing smallholdings to shareholders chosen by ballot. The difficulties lay in the nature of the subscribers, mainly skilled townsmen with little aptitude for farming; the hostility towards the Chartist leader, Fergus O'Connor; and the mismanagement of the company's financial affairs which led to a parliamentary enquiry. Significantly, the Select Committee found no evidence to justify rumours of fraud, but confidence was sufficiently dented for the flow of share capital to dry up. O'Connor followed the advice of the parliamentary committee and wound up the company after only five years. We can see that erratic management clearly contributed to the downfall of the Land Plan, but equally parliament made no effort to keep it alive. The weakness of the scheme lay in its origins, its isolation from mainstream farming, and the opposition of the ruling elite. Nevertheless, the movement was sufficiently strong to attract huge interest in small-scale farming, which continued after its demise.[23] Much more influential amongst the moderate Liberals and the middle classes was the writing of the political economist J.S. Mill, who reviewed the question of land reform in his magisterial work, *Principles of Political Economy*, in 1848.[24]

Mill was the first English economist to write favourably on *métayage*, as part of a critique of the large heavily capitalized farms which dominated English agriculture by the mid-nineteenth century. His principal aim was to highlight the benefits of small farms. Mill fully appreciated the disadvantages of *métayage*, as articulated by Adam Smith, but was critical of Arthur Young, whose 'unmeasured vituperation … was grounded on an extremely narrow view of the subject'. His focus on France on the eve of the Revolution was misleading, 'the situation of French métayers under the old regime by no means represents the typical form of the contract'. [25] For the system to work effectively, it was essential that the proprietor paid all the taxes. But in France the exemption of the nobility from direct taxation had led the government to throw the whole burden of taxation upon the occupiers, and it was to these exactions that Turgot ascribed the extreme wretchedness of the *métayers*.[26] To refute Young's argument, Mill cited the case of *métayers* in Italy, where the system worked well and produced enormous social benefits.

Mill sketched a Utopian scene in Italy. In Lombardy and Piedmont, the farms of between 10 and 50 acres were all occupied by *métayers* at half profit. The farms possessed a 'richness of buildings rarely known in any other country in Europe … the rotation of crops was excellent … no country can bring so large a portion of its produce to market as Piedmont … In no part of the world are the economy and the management of the land better understood than in Piedmont, and this explains the phenomenon of its great populations and its immense exports'.[27] All this was achieved by *métayer* cultivation. In the Arno Valley, around Florence, forests of olive trees covered the lower slopes, hiding an infinite number of small farms of between three and ten acres, with houses surrounded by orange groves and vineyards, and the land between divided by canals or rows of mulberry trees and poplars, the leaves of which were eaten by cattle. The farms maintained a horse and cart, and two oxen for the plough, creating an appearance of comfort and prosperity. The reality was rather different. Within the Florence area *métayers* could supplement their income with domestic crafts, such as strawplaiting, but beyond Florence, owners had to provide half the stock, and lend the *métayers* money to buy the other half, sometimes even money for corn to eat. To be fair, in this area of tiny farms, the comparison was best made with English labourers, rather than farmers, and from that perspective, the view of commentators was generally favourable. The rich proprietors enjoyed a close relationship with their *métayers*, concerning themselves with every aspect of their activities unlike the great landowners who leased their farms at fixed rents. These Florentine landlords, interested in quality and diversity, never refused an advance, paid for irrigation systems, terracing, and all those gradual but permanent improvements which could not be financed by the *métayers*. Thus together they perfected the rural economy of Italy, producing a great range of crops and small scale luxury products for export and the table, including wine, olive oil, fruit and grain. Above all, the relationship fostered a community of interest, a kindness between proprietor and *métayer*, which benefited the moral condition of society.

Significantly, the partnership between proprietor and *métayer* was governed by usage or custom, rather than contract. The *métayers* treated their *métairies* as an inheritance; they knew every detail and planned for the future. Although he did not own his farm, in practice, the *métayer* enjoyed optimal conditions for holding land: a permanent tenure paying a fixed proportion of his crop, while benefiting from the financial support of the proprietor, who paid the taxes and invested in capital improvements. In contrast, the English tenant farmer knew he was a temporary occupier holding his farm by a contract for a term of years at a negotiated rent, which forced him to maximize his profits. We can see that *métayage* particularly suited the production of high quality, risky products, such as olives and vines, which required substantial capital investment at the outset, and the most careful cultivation. Farming of this kind needed long-term commitment and a sensitive partnership between the parties. The landlord-tenant system with its rigid distinctions was not conducive to this type of enterprise; as we have seen the production of quality cheese in Dorset and Wiltshire required a different relationship, involving the provision of working capital to the tenant.

Mill did not advocate the adoption of *métayage* in England on the grounds that it was not an indigenous practice. In his view a system based on custom needed to be rooted in the culture of that society. However, he warned against the abolition of *métayage* in countries where it was well established 'based on an *a priori* view of its disadvantages'. If the system worked in Tuscany, let it be. He objected to the French economists who tried to replace *métayers* with farmers paying money rents; it was better to encourage them to build up stock than replace them.[28] In this, Mill anticipated the revival of *métayage* in France in the depressed conditions of the 1880s.

In his chapters on the virtues of peasant proprietorship in various parts of Europe, Mill described the manufacture of Gruyère cheese in Switzerland.[29] Throughout Europe cheesemaking involved intricate arrangements with skilled workers. Each parish hired a man, from the district of Gruyère, to take care of the herd and make the cheese; he worked with a pressman, or an assistant and a cowherd. The average herd consisted of about 40 cows, with parishioners each contributing two or three beasts. The cheeseman and his assistants recorded the yield of each cow and then put all the milk together and made the cheese. Mill marvelled at the trust placed in the workmen. Each owner received the weight of cheese in proportion to the quantity of milk produced by their cows. By this co-operative method large high quality cheeses were produced by specialists for the market. The staff were paid so much per head of the cows in money or in cheese, sometimes they hired the cows and paid the owner in money or cheese. A similar system existed in the Jura, and reminds us of arrangements in Dorset and Wiltshire.[30] We can see how this system addressed the twin problems of all small producers; quality control and marketing. A crucial aspect of the system was the incentives it provided for the dairymen. For similar reasons, Mill argued that peasant proprietorship, by giving them a stake in the land, would motivate labour, raise production per acre, create a great diversity of crops for the market and develop the intelligence of the peasantry. Sentiments such as these informed the debate on land reform in the last quarter of the nineteenth century, as agriculture slid into depression. Ironically, through his ignorance of the practice in England, Mill more or less closed the door on *métayage*, the revival of which was proving so successful in France.

The debate on land reform in the first half of the nineteenth century was not confined to the plight of the landless labourer. Of possibly more immediate concern to landowners and tenants was the question of the right of tenants to compensation for unexhausted improvements on the surrender of a farm, indicating still further the growing tensions within the landlord-tenant system.[31] Tenant right assumed great significance after 1815. First, the fall in agricultural prices and reduced profits led to a greater turnover of tenants and second, from the 1830s, the purchase by tenants of off-farm inputs, such as artificial fertilizers and stock feed, needed to be recognized. The failure to secure the right of compensation for these inputs deterred tenants from investing capital during the final years of their lease, adversely affecting the condition of the farm for the incoming tenant and setting back the course of improvement. In these circumstances long leases, thought to be a prerequisite to improvement, became a hindrance rather than a help. A solution emerged in Lincolnshire, a county noted for its lack of leases and flexible systems of tenure, and for the prevalence of *métayage* in the seventeenth century.[32] Known as the Lincolnshire Custom it later formed the basis for the new law of tenant right embraced by the Agricultural

Holdings Act of 1875 and 1883.[33] The custom developed after 1815 to compensate tenants' improvements on the Lincolnshire Wolds which had been brought into arable cultivation in the late eighteenth century. The fragility of the farming regimes on these light poor soils necessitated the development of a strategy whereby the outgoing tenant would hand over the farm intact to the incoming tenant. On these carefully and expensively improved farms the cost of permanent damage could not be tolerated. In essence the custom amounted to a partial compromise of the principle that whatever is put into the soil passes into the soil, and benefits the landlord at the expense of the departing tenant. Instead, a limited tenant right granted compensation for inputs that increased the fertility of the soil, but were not exhausted during the period of tenure. In calculating the compensation, time periods were established over which the cost of the tenant's inputs was progressively discounted by a certain proportion each year until the fertilizing value was assumed to be exhausted for the purposes of tenant right. Thus the cost of an application of bone dust to the turnip crop on a Lindsey farm was usually discounted by one-third for each year of a three year period until the right to compensation was exhausted. On quitting his farm a tenant was entitled to reimbursement for the remaining proportion of his outlays on fertilizing inputs according to the time-periods established under the custom for each particular input.

Other systems developed to tackle the problem of tenant right, but the Lincolnshire Custom proved to be more flexible and effective than most. This was due to two factors. First, the compensation was based upon the actual outlay made by the tenant, rather than the value to the incoming tenant. Second, compensation was assessed on the number of years needed to recompense the outgoing tenant for his outlay, rather than the length of time required to exhaust the improvement. For example, marling on the wold hills was discounted over a period of seven years, although the task was only repeated every twenty-five years. In this way a balance was struck between the interests of the outgoing and incoming tenant, an issue which became particularly important during the Great Depression. What appears significant in Lincolnshire is that the power of custom proved more efficient in the changing circumstances and less certain conditions of the nineteenth century, than the long leases advocated in the eighteenth century. This flexibility and respect for custom may be linked to systems like *métayage* which had been common in this remote and sometimes difficult county, where collaboration and trust between landowners and farmers was essential. As we saw, Massingberd drew attention to the prevalence of *métayage* in Lincolnshire during the seventeenth century.[34] Writing in the 1890s, he may have been recommending its return as a solution for Lincolnshire landowners and farmers in those depressed times.

The discussion of tenant right, alongside the plight of agricultural labourers, draws attention to the inherent difficulties within the English landlord-tenant system, which had become apparent long before the great depression in agricultural prices. Despite the calls for land reform, from radicals and moderate liberals led by Mill, the landed elite offered no more than half-hearted palliatives in the form of allotments. In the second half of the nineteenth century, the deepening depression required landowners and the government to be more pro-active, but success was limited.

The need for flexible strategies and alternative solutions became paramount in the late nineteenth century as English agriculture entered a period of long-term decline. From the late 1870s, agricultural prices slumped as the imports of cheap corn, prophesied by the landed interest during the debate on the repeal of the Corn Laws, finally arrived from Canada and the USA. This was followed by refrigerated meat and dairy products from Argentina, Australia and New Zealand, and more ominously, inexpensive and high quality butter, bacon and cheeses from France, Denmark and the Netherlands. Only milk, hay and straw escaped foreign competition.[35] Globalization on this scale required English farmers to diversify from corn into milk, beef, and small scale enterprises such as poultry, market gardening and horticulture; types of farming that were successfully undertaken by small farmers. Ideas for the promotion of smallholdings, which had been ridiculed in the 1830s and 1840s, became mainstream politics by the 1870s, when Joseph Chamberlain and Jesse Collings launched the Unauthorized Programme advocating the provision of 'three acres and a cow' for rural labourers.[36] Such schemes acquired a new urgency as the more enterprising left the land for better opportunities in the towns or emigrated to the colonies. How to stem 'the drift from the land' became a principal concern amongst the informed classes. Reformers produced a raft of proposals, from allotments to communes, cottage farming to profit-sharing, peasant proprietorship to *métayage*. Many schemes proved hopelessly impractical, for, as the economic situation deteriorated, so land reform became enmeshed with nostalgia for an idyllic rural past. As England urbanized and people lived in dirty cities, literature, songs and paintings harked back to a period of lost content, when the land was cultivated by yeomen farmers and labourers grazed their cattle on the commons. Soon men and women were being actively encouraged to forsake the grim realities of town life and move back to the purity of the land.[37] The enfranchising of labourers in 1885 also required a response. Both political parties recognized the need – and the mood – and passed new laws extending the provision of allotments in 1882, 1887 and 1890, while the Small Agricultural Holdings Act of 1892 gave the new County Councils formal powers to purchase land to sell or let to smallholders.

The literature from this period, while not straightforward, provides insights into prevailing attitudes to land tenure and the interest in profit-sharing and co-operative schemes. The rigidity of the 'tripartite system' of landowners, tenants and labourers was a common theme. The perception amongst land reformers was that it worked against the interests of the rural community and made the necessary changes to English agriculture extremely difficult to effect. Broderick, a leading protagonist, argued that the benefits of the landlord-tenant system were largely theoretical. In practice, estates were plagued by absenteeism, drained of profits and starved of investment, with outgoing tenants unable to secure compensation for their improvements. Added to this, the strict family settlement ensured that huge estates remained in the hands of single families, resistant to change and often heavily indebted. These issues were addressed in 1882, with the Settled Land Act, which facilitated the sale of entailed land, while the Agricultural Holdings Act of 1883 granted the principle of tenant right for unexhausted improvements.[38] Like so many commentators, Broderick was strongly influenced by progress on the Continent, notably France and Denmark where small farms and 'la petite culture' proliferated. For England he advocated a system of cottage farming and 'the revival of old *métayage* tenure' for holdings of between 20 and 30 acres. He defined 'the essence of this ancient tenure, which seems to have once prevailed in many parts of England', as the provision of farming capital by the landlord, not the payment of rent in kind, or the equal division of the profit between landlord and tenant.[39] This is not entirely correct, but like Thorold Rogers, he compared the system to the medieval stock and land leases, which resembled the *métairies* of south west Europe.[40] In Broderick's view, a modified form of *métayage* offered a solution to the question as to whether the current system supporting landlords' rent, tenants' profits and labourers' wages could be maintained with the continuing fall in prices. In times of adversity, the cottager enjoyed an enormous advantage with his low costs and small-scale enterprises; he made better profits and paid higher rents per acre. To help him market his produce, Broderick advocated co-operative agriculture. Finally, he foresaw the growth of a new class of small freeholders, frequently local tradespeople, miners and petty rural landlords with access to capital.

However, like Mill he was not optimistic about the prospects for the revival of true peasant proprietorship in England. The best opportunities existed in the colonies, Canada, New Zealand and Australia.

Writing 20 years later, Shaw-Lefevre reviewed the progress of land reform and identified the intractable problems that continued to prevail.[41] Like Offer he believed the root cause of these problems was the political power vested in the landowning classes.[42] The prestige, status and power associated with the ownership of land made it the ambition of men of fortune, raising the value of land and putting it beyond the reach of those who looked upon it as a mere investment. Enclosure had exacerbated the process, robbing smallholders of common land on which they depended, and leading to the near extinction of small landowners faced with the prohibitive costs of the enclosure. He cited the example of a merchant in Westmoreland accumulating an estate from 226 statesmen. The aggregation of large estates had led to large tenanted farms, economies of scale, fewer buildings and larger machines, but almost eliminated the culture of small-scale production. Small tenancies and farms could only survive on the fringes of towns where farmers could supplement their income by milk and egg rounds, carrying on a trade or by offering a service. He believed that agricultural statistics grossly exaggerated the number of small farmers. Of the 236,000 persons listed in 1875 owning holdings of less than 20 acres, there was no information on their actual occupation. Were these peasant proprietors or gentlemen owning villas with a small park attached? He gave an example of his parish in Kent where the land was fairly evenly distributed and the growth of fruit for the London market encouraged smallholders. But there was not one case of a peasant making a living from it, most were tradespeople cultivating a few acres as a sideline, just as Broderick had predicted. Tenancies of between 40 and 100 acres still survived in Yorkshire, Lancashire, Devon and Wales, but in most parts of the country they had vanished.[43]

From our perspective, it is significant that Shaw-Lefevre was a Liberal MP, steeped in the ideas of J.S. Mill, and a founder member of the Commons Preservation Society. He was also chairman of the Select Committee on the Irish Land Act, 1875, and later chairman of the Royal Commission investigating the agricultural depression. His advocacy of peasant proprietorship frequently annoyed Gladstone, who was intent on wider tenurial reform, particularly in Ireland. As part of the political establishment, Shaw-Lefevre accepted that in many ways the tripartite system had produced excellent results: improved and well equipped farms, good cottages, allotments, moderate rents and generous abatements. Tenants and labourers felt confident in this supportive framework, and preferred to lease big farms than buy small ones and undertake the responsibility for the entire fixed and working capital. The system had become established practice and embedded in the culture of most rural communities. However, in depressed times, the system proved cumbersome and inflexible, and this aspect he sought to rectify. Like Broderick, he agreed that small farmers survived better in these circumstances than their large neighbours. With greater flexibility and lower costs they could turn their hand to profitable small-scale enterprises, poultry, dairies, fruit and vegetables. By the 1890s the neglect of this social group had proved extremely costly to English agriculture. The Small Holdings Act of 1892 tried to tackle the problem by creating small ownerships and a farming ladder for enterprising labourers, under the sponsorship of local authorities, who were empowered to purchase land and either sell or let it to smallholders. Both political parties supported the sentiment, although the Conservatives were more sceptical of the viability of the scheme. The holdings would be large enough to support a living, and most importantly for their success, tradesmen were not excluded from the scheme. Yet, from the outset, the cost of dividing farms and equipping the smallholdings with buildings, fencing and water, proved to be the stumbling block. Moreover, agricultural labourers simply did not possess the capital to stock and farm these holdings, and possibly not even the skills required to earn a living from 30 acres. 'La petite culture' was largely absent in England, and as Mill predicted, such arts proved difficult to instill. Most Englishmen by that time preferred the safety and simplicity of hiring a farm, or working for wages with perquisites. The vast majority of holdings were taken, as predicted, by tradesmen who needed grazing for a horse, or wanted to keep a cow, but this was hardly the purpose of the Act.[44]

The experience of Lord Wantage, who founded the Small Farms and Labourers Land Company in 1885, shows how far reality differed from the sentiment.[45] A renowned agriculturist and land reformer, he promoted the purchase of estates for breaking up and selling to smallholders, and had himself donated 400 acres of the Lambourn estate in Berkshire to the company. A Tory MP for many years, he was motivated by the desire to create a body of men interested in maintaining the rights of property, as an antidote to rising Trade Unionism. He was related by marriage to Shaw-Lefevre, who cited his efforts fully in the Berkshire report for the Royal Commission of 1893–4.[46] However, despite heavy subsidies, the scheme did not succeed. First, Wantage found it difficult to secure any buyers. Selling land to labourers proved a dead letter, but letting was another matter, particularly if tenants were allowed to combine farming with another activity. Farm labourers could only afford to stock and cultivate about five acres, so the larger parcels were invariably let to village tradesmen, who had access to other sources of capital and were not wholly dependent on agriculture. A more successful scheme was initiated by Sir Thomas Acland in Cornwall. He divided and leased a moor to tin miners, allowed them a few acres and a place to build a house, and granted them security of tenure. This echoes Mill's belief that permanent tenure, combined with varied employment, was the key to success for *métayers* and small farmers. Despite these examples, the concept promoted by the reformers was largely alien to the outlook of landowners; they had no desire to create colonies of independent small owner-occupiers close to their estates. In reality, it proved almost impossible to turn the clock back.[47]

In the 1890s, the idea of profit-sharing in agriculture received a considerable boost through the agency of Earl Grey, who had, through his ginger group, the Grey Committee, long promoted industrial co-operatives and profit-sharing.[48] In 1884, he took over the management of his uncle's large estate in Northumberland and applied these principles to agriculture. In an article written for the Royal Agricultural Society in 1891, he encouraged landowners to follow the same route. Grey was also related by marriage to Shaw-Lefevre, who publicized his experiment in the Northumberland Report of the Royal Commission, complete with supporting tables of Grey's success on the profit-sharing farms at East Learmouth (our Type B).[49] Grey believed that profit-sharing offered a better hope of strengthening the relations between employer and employed, of remedying agricultural discontent, and of bringing about a moral and material improvement in the individual circumstances of those engaged in the production of wealth, than any other principle which had been before the public. Profit-sharing was a very potent means for the improvement of agriculture in this country. In this we detect a strong political message, the purpose being to promote profit-sharing as an alternative to the schemes for peasant proprietorship advocated by Mill and the Liberal establishment.

In Grey's view, profit-sharing was preferable to peasant proprietorship as it united the advantages of a motivated workforce with the economic benefits of large farms. This explains why it won significant support from parliamentarians, including Gladstone and Balfour. Furthermore, the scheme did not challenge the existing social structure; landowners retained a controlling role as arbitrators, were able to discriminate and exclude members from the co-operative and would decide on the annual bonus. The schemes varied in their detail, but essentially the landowner obtained a rent for his land and interest on the working capital, stock, implements he provided, while the labourers were paid their usual wage. At the end of the year the profits were shared between the landowner, bailiff and labourers, according to a formula devised by the landowner. Grey used several examples to demonstrate the efficacy of the system and yet it seems that, just as in the depressed conditions of the late seventeenth century, these unconventional schemes were confined to enlightened landowners.[50] He pinpointed two aspects which deterred the average and the suspicious. First, the system required extremely accurate book keeping, and secondly, tenants were reluctant to join the schemes and reveal their profits for fear of their landlords raising rents. The position of landlords, as effectively a third party to the agreement, certainly made such arrangements more difficult to implement than a straightforward relationship between landowner and labourer. Grey's own motivation for the adoption of profit-sharing on his Northumbrian estate stemmed from the difficulty of retaining the young male labourers in the face of growing competition from industry. Shortage of labour was a long-standing problem, and Northumbrian landowners were used to offering perquisites, including the right to keep a cow: 'The farmer is ready at the hind's request, to provide a year's keep and the necessary accommodation for a cow, in return for the payment of 3s or 4s a week. For this sum the farmer allows the cow to run out to pasture in the summer, and in winter he gives the labourer a sufficient allowance of artificial food. The advantage to the labourer of this arrangement is greater than may appear at first sight. An average cow will give the labourer a gross income of 5s to 6s a week, or even more, and over and above this, the hind has skim milk for his own use. He has, therefore, every inducement to try and obtain possession of a cow'. This scheme we recognize as a form of cow leasing.[51]

Building on these collaborative relationships, Grey set up his profit-sharing farm at East Learmouth, which involved the cultivation of 821 acres, including 122 acres of pasture. With adjustments, payments into a reserve fund and additional rewards for the bailiff, the workforce could expect about 1s in the £ every year; in other words a 5 per cent increase on their wages. It is important to stress that Grey's profit-sharing scheme was designed to provide incentives to wage labourers, he did not advocate farming to halves with a tenant.[52] However, he was familiar with the practice and cited the example of Lady Carlisle who used the *métayage* system on the Castle Howard estate in Yorkshire. The tenants paid just a third of their sales to Lady Carlisle, which she stated amply covered the rent and interest on her working capital. Grey explained the system known as *métayage* and the attempts in France to revive and modernise it over the previous 20 or 30 years.[53] Attacked in France by the advocates of fixed rent tenancies, *métayage* found enthusiastic defenders, even amongst the most illustrious agriculturists. He quoted several publications, including the work of Count de Gasparin on the *Rural Economy of France* (1830), and selected references pointing to its success in France from Britanny to the Beaujolais vineyards; from the arable cultivation of the Dombes (Ain) to the pastures of Limousin. It was commended for its certain elasticity and application in the most diverse circumstances and localities.[54]

At the same time as Grey was writing, Betham-Edwards, in her introduction to Arthur Young's *Travels Through France* and in her own book, *France of Today*, directly attacked Young for his condemnation of *métayage*. 'Nothing has done more to improve French husbandry within the last fifty years', she averred.[55] In contrast to Britain, French agriculture had advanced without sacrificing the well-being of the labourer. She described the *métairie* of 400 acres in Berry, belonging to her host. 'The system is a partnership, the owner supplying land rent free, stock and implements; the *métayer* manual labour, all profits being equally shared … The *métayer* boards his farm labourers, as was the custom in England fifty years ago', and the labourers' wages, she reckoned, had risen fivefold in the previous 20 to 30 years. *Métayage* effectively harnessed the labour and energies of all parties involved, providing incentives and goals. 'It is the stepping stone from the status of hired labourer to that of capitalist, and whilst the *métayer* is thus raised in the social scale, by his means vast tracts are brought under cultivation'. *Métayage* was used to bring millions of acres of waste and heathland into cultivation.[56] Writing a few years later, Henry Higgs also commended the achievements of *métayers* in Western France, and speculated on the possibility of introducing *métayage* to England. He drew attention to the social obstacles, and doubted whether English owners or farmers would 'take kindly' to the system; the first would dislike the bother of supervision and the uncertainty of income, while the latter would resent the interference in their farm management.[57] However, he noted that as English farming continued to decline, and owners and farmers were forced to consider alternatives like profit-sharing, 'we shall not be far removed from the contract of *métayage*'.

Writing today, Carmona provides an example of the French revival, showing how landowners in Poitou, Berry, Limousin and the Bourbannais, areas long associated with *métayage*, adopted sharecropping to develop new livestock enterprises from the 1830s to the 1930s.[58] They favoured sharecropping contracts, as opposed to fixed rent tenancies, for a number of reasons. In these remote and underdeveloped areas, well capitalized tenants, capable of taking on large farms, did not exist. But landowners needed large holdings to achieve economies of scale in land clearance, the construction of new roads and buildings, and the breeding and marketing of livestock. Sharecropping contracts allowed them to keep control of the fledgling enterprise and protect their capital investment. Landowners also needed skilled labour to care for the livestock and look after the investment. As they received a proportion of the uplift on the cattle, sharecroppers were more motivated and required less supervision than wage labour. As with cow leasing, sharecroppers used family labour in the most effective ways, rendering it much more efficient than wage labour. In this context, sharecropping was the optimum solution for landowners. In areas lacking in capital and human resources, where wage labour was often expensive and inadequate to the task, sharecropping was a valuable alternative. As we saw at Raynham in 1679, Timothy Felton recommended 'putting to halves' as the cheapest method of effecting improvement to the pastures in the park.[59]

In her *Alternative Agriculture*, Thirsk draws attention to the literature on profit-sharing, noting that in the 1880s and 1890s, it was very much a live issue as farmers sought solutions to the agricultural depression. Even Prothero, that arch defender of large farms, extolled the virtues of *métayage*, 'it has in France proved the best shelter against the agricultural storm'.[60] And yet, despite contemporary interest, historians have tended to overlook these ideas, viewing land reform simply through the prism of allotment provision and the creation of smallholdings.[61] Freeman, in his work on social investigation in rural England, accepts that the land question was the most important factor determining social conditions in the countryside. Yet, he does not pursue the point, beyond characterizing a landed society divided between those who defended the status quo and those who blamed the land monopoly as the root cause of social ills. Social investigation – and land reform, such as it was – should be understood within this context of vested interests and deeply polarized positions. He makes no reference to profit-sharing and co-operation across the social divide.[62]

More surprisingly, neither does Mills, who argues for the survival of a strong peasant tradition in rural Britain comparable to the European model.[63] He distinguished a 'peasant system' in open villages in contrast to the 'estate system' occupying the neat but socially suffocating closed villages. In these places, the familiar struggles took place, with men such as Joseph Ashby leading the battle against Lord Compton to secure allotment land for the villagers of Tysoe. Mills cites the example of Lord Wantage as a paternalist Tory land reformer and makes no mention of his profit-sharing schemes. Like Freeman, he stresses the aspect of social control, rather than collaboration, reflecting his dichotomy between estate and peasant systems. In his peasant villages, Mills also overlooks the possibility of collaboration between tenants, despite the fact that it was difficult to ascertain from the records who worked for whom and the actual occupancy of holdings. A close study of a peasant village showed that many small properties were not occupied by their owners, but by shopkeepers and retired men and women living on their savings. Thus, following the Chayanovian model, a small farmer could build up a larger holding as his family capital and labour expanded, by renting fields and homesteads scattered across the parish. These circumstances provided a congenial environment for farming to halves but Mills does not refer to the practice or consider the possibility of forms of *métayage*, which is strange given that he is attempting to draw similarities between the English and Continental experience. From these accounts, we can see that English historians are not generally receptive to ideas of co-operation and partnership, and simply do not regard them as relevant in the English context. Lack of documentation has played a major part in shaping this outlook, but so has a preoccupation with other concerns. As so often in this hidden history, modern historians reiterate themes that emphasize differences, ignoring strands that link us more closely to our counterparts in Europe and the New World.

Despite the efforts of reformers, alternative solutions, such as profit-sharing and *métayage* remained a minority interest in England. As we have seen, economists and agriculturists, from Mill to Broderick held out little hope that their ideas would be taken up by English landowners, firmly in the grip of the tripartite landlord-tenant system. The best prospects for English labourers and smallholders, eager to advance themselves, lay in the far flung colonies of Canada, Australia and New Zealand. The traditional view is that the modern interest in sharefarming stemmed from the New Zealand experience. Less appreciated is the impact of British land reformers on the colonists in the late nineteenth century. These peasant settlers, determined to resist any imposition of a landlord system, proved most receptive to land tenure reform championed by Mill and took to New Zealand practices long familiar to them in England, Wales and Scotland. In this way, sharefarming was exported to the colonies, only to return to England in the 1980s as the culture of landowners changed in response to new legal and financial constraints.

The success of New Zealand's agriculture today blinds us to the problems encountered by the colonists at the outset.[64] They faced all the difficulties of a pioneer economy: virgin lands, lack of capital, poor infrastructure, limited markets and scarce labour and, in addition, extreme isolation. European settler farming conformed to two types: the large sheep stations which relied heavily on Australian stockmen and secondly, family farmers employing no hired labour, who scratched a living from dairying and horticulture, supplemented by contracting, haymaking and fencing. There was no equivalent of the English farm, owned by a gentleman, farmed by a tenant, and worked by wage labourers. In this land of scarce human resources, wages were high.[65] The early attempt to export an English landlord and tenant system soon foundered. Gentry colonists, appalled by the levelling down that they encountered, returned home, while workers refused to accept a servile status. As Sinclair wrote, they delighted in a forced equality.[66] By the 1880s the old establishment of Wakefield settlers had fallen apart, their fate sealed by the influx of immigrants with the goldrush, the introduction of refrigeration in 1882, and the Liberal victory of 1892. In that election, the political initiative passed irretrievably to the small farmer and the common man. The New Zealand Liberals, Fabian in outlook and strongly influenced by J.S. Mill, Henry George and A.R. Wallace, imposed a land tax on the run-holders and subsidized the division and redistribution of the large estates.[67] At the same time, the government provided cheap credit to smallholders to buy land and stock. In this egalitarian and collectivist society, co-operatives and sharefarming found ready acceptance.

The advent of refrigeration in the 1880s transformed New Zealand, economically, socially and politically.[68] The overwhelming beneficiary was the small family farmer. From the 1890s, dairy farmers could export their butter and cheese, as well as meat products. Between 1896 and 1914 dairying activities and output increased five times, while cheese production rose ten fold. By 1926, 60 per cent of farms were family farms, run with family labour. This was reflected politically as 'Cow Cockies' became a huge force in the Farmers' Union and campaigned for further land reform and more cheap credit. Their particular concern was to secure the freehold of their farms, which they preferred to the perpetual leasehold considered by many as the backdoor to land nationalization. Only high land values, accompanied by cheap credit, provided sufficient incentive to these pioneers breaking in and working new farms. Such was their political clout, that the government responded with a raft of measures to assist small farmers. They set up co-operative marketing systems, education and research institutions, disease and quality controls, and built the infrastructure of roads and rail essential to transport agricultural produce. Most significantly, they established a national rural bank, which guaranteed cheap loans to young men eager to purchase stock and land. Thus the family farmer was subsidised three times over: with the purchase of his stock, his land, and his labour costs when he employed a sharemilker. All were supported by cheap credit. In these ways, the government generously underwrote and shared the risk of farming. The political muscle of the small farmers ensured that this level of support continued until the mid-1980s, when subsidies were gradually phased out.[69] Yet even the removal of price support and protective tariffs did not strike the death knell of family farms. 'Cow Cockies', in particular, proved amazingly resilient; they found new markets, developed new products, and cut already low overheads to the bone. The Review of Sharemilking in 1996, showed how they had successfully weathered fluctuating incomes and low returns, providing a lesson for their subsidy-bloated European counterparts.[70] Increasingly resourceful, New Zealand dairy farmers now produce 30 per cent of the world trade in dairy products without any form of subsidy.

Within the structure of the family farm the key to success was, and remains, sharemilking which provides the industry with a steady flow of young and enterprising new entrants. Of the dairy farms in New Zealand, 28 per cent are operated by sharemilkers and their production is 15 per cent higher than that in owner-occupied farms. The main reason is that the average age of the sharemilker is 33, an age at which he is at his most productive, enabling him to optimize the size of his herd at between 200 and 250 cows. On the other hand, a small owner-occupier, with capital invested in land, might have to limit himself to a 100 head herd. A typical New Zealand sharemilking contract is similar in principle to farming to halves or *métayage*. The farmer provides the land and buildings, the 'operator' the cows, labour and machinery, and they divide the monthly milk cheque on a 50:50 basis. Where the operator does not provide the herd, the proportion is reduced to 39 per cent for the use of his labour and machinery. Alternatively, he or she can milk the cows on a contractual basis, with payments and incentives for achieving specified production levels. Less common is the 29 per cent sharemilker with neither machinery nor cows; he is invariably better off working for wages. The lifecycle of a New Zealand dairy farmer follows the Chayanovian model, with his holding expanding and contracting as his needs dictate and abilities permit. In his early twenties, he builds up by contract milking enough equity to pay a deposit on a herd of 100 cows; he then becomes a 50:50 sharemilker, servicing his cheap loan with the Rural Bank. In time, he progresses to a bigger farm to support a much larger herd.[71] He then decides whether to continue sharemilking, with an average return on investment of between 20 and 30 per cent, or to sell half his herd to finance the purchase of a small farm. As land prices have increased since the 1980s, more sharemilkers opt for the former, yet the survey of 1996 showed that 53 per cent still aimed to buy their own farm. The cycle continues into middle age; he may in his forties, choose to buy a larger farm, or retire and take on a sharemilker. Those who remain sharemilkers frequently diversify into other forms of investment: shops, businesses, residential property. Thus the profits of agriculture are pumped into the wider rural economy. The opportunity to build equity, which sharemilking provides, is the key; it not only guarantees a highly motivated workforce, but ensures a steady turnover of farms as ageing owners retire and sell to sharemilkers. Without acquiring equity in this form, old farmers would be locked into increasingly inefficient production, as they are in Britain, where the average age of farmers is 58 years. These rigidities in British dairy farming encouraged enterprising young men in the sector to try their luck in New Zealand. They returned often asking the question, 'why not sharemilking in Britain'?[72]

Why has sharefarming in New Zealand proved so successful? New Zealanders will tell you that much can be attributed to the climate as grass grows all the year round, albeit at different rates. Thus inputs are low and easily managed, keeping down costs and helping to protect the profit margins of both parties. But the institutional structure has played a major part, not only in providing cheap credit and a legal framework, but also especially in the marketing of the produce through co-operatives. The dairy companies provide economies of scale to farmers and control the payment to the farmer and operator, reducing the likelihood of dispute. The monthly account, itemizing the milk produced by the farmer, overcomes the traditional difficulty of monitoring dairy produce, collected every day and sold in different forms. In France, sharecroppers were restricted in their sales of butter and cheese, until the development of co-operatives in the 1890s. In this way, the co-operative, by performing the role of monitor, significantly reduced transaction costs for landowners and operators. This helps to explain why the English, when they started to create commercial dairy herds in the late seventeenth century, resorted to cow leasing, where the dairyman received the entire produce, rather than attempt a system of farming to halves. The New Zealand dairy industry was most fortunate in its timing, benefiting from modern technology, land reform and a political and social culture which fostered flexible and enterprising family farms.

The nineteenth century is the only century for which we have no archival evidence of farming to halves in England. Its reappearance in farm diaries in the early twentieth century suggests that this may be an anomaly and that further research would uncover examples of it. Most of the written evidence for farming to halves derives from estate records. When it ceased to be a practice favoured by landowners in the early eighteenth century, it disappeared from the documents, while examples of the practice further down the social scale have never been easy to locate. Inevitably this chapter has been skewed towards well documented schemes advocated and implemented by landowners to relieve the conditions of their landless labourers. The provision of allotments, profit-sharing and the encouragement of smallholdings all reflect an appreciation of their loss, and their need to be involved and rewarded for the land they worked, but none of these strategies amounted to sharefarming in the strict sense of the word.

However, from the writings of Reed and others, we sense that farming to halves continued in the nineteenth century amongst peasant circles. The survival of these groups in significant numbers, and the ways in which they organized their working lives, indicate that forms of sharefarming played an important role in rural communities, providing credit, employment and support systems, just as they had always done. Further evidence can be deduced from the interest in schemes, such as the Chartist Land Plan, Chamberlain's Unauthorised Programme and the popularity of sharefarming in the English colonies. The success of sharemilking in New Zealand points to the hidden English traditions of cow leasing and sowing to halves. Without the cultural constraints, and the stifling grip of the English landlord-tenant system, sharefarming in the colonies flourished. As attitudes changed in the late twentieth century, sharefarming became more acceptable to landowners in Britain. As we shall see, in peasant circles, it had never entirely disappeared.

8
A Return to Halves in the Twentieth Century

In the early twentieth century, farming at halves reappears in documents after an absence of nearly 200 years. This is not profit-sharing, associating labourers with philanthropic landowners as in the nineteenth century (our Type C), or the cow leasing of the eighteenth century (Type G), but a return to the classic version of sharefarming encountered in the early modern period, where landowners and farmers agreed to divide the produce, rather than have a fixed money rent (Type A).[1] The evidence, recorded in farm diaries and newspapers, strengthens our belief that the practice never died out amongst farming families. Yet, as in the nineteenth century, no references to farming to halves in this period have been found in contemporary or modern academic accounts. Indeed, in 1923, Venn commented that, despite the advantages *métayage* had conferred on French agriculture, the practice could never have flourished in England; 'it calls for a too domestic partnership between landlord and tenant, a partnership as uncongenial to the type of landowner found in this country, as to his more independent tenant'.[2]

But for a chance discussion with present-day working farmers with long memories, we would still have no idea that the practice had survived into modern times. This encounter led to the discovery of two farm diaries which showed, quite clearly, that placing ewes at halves was a regular practice in south Shropshire from the 1900s to the 1930s. Much later a paper given to the Bishop's Castle Quality Cattle Association led to another farm diary and a recommendation that we consult advertisements in the local papers. Admittedly, a limited sample, but it showed that the practice of putting out ewes at halves extended from Shropshire through Herefordshire to Monmouthshire, and from Worcestershire into Wales, throughout the 1920s and into the 1950s, in precisely the same way as it had existed in this area in late medieval times.

This chapter begins with an analysis of the agreements from Shropshire. The reason for the concentration here is not that farming to halves was confined to that county, but that one of the authors, Elizabeth Griffiths, had close links to several families who have farmed in the area for generations. Similar connections in Cornwall, Cumbria or Devon might well have yielded similar results. Membership of this farming community gave unique access to privately held diaries, papers and photographs. The Museum of Rural Life collected many farm diaries in the 1950s and might have been expected to yield more recent evidence of farming to halves, but it seems that the most precious private items about farming agreements survive in oak sideboards, carefully preserved as family heirlooms, brought out for inspection by trusted visitors. These were not dusty books stored in the attic, but regularly consulted for details on family history and insights into ways of farming. Just as William Windham kept his Green Book, these records were created as reference works for future generations. The families all recognized that their diaries contained references to farming at halves, and were interested to learn more of its wider historical significance. Once alerted to this, they were able to adduce more evidence, offer fresh insights and suggest new lines of research, which included looking at the advertisements in the local papers. The recurring theme from these discussions was the privacy which surrounded the conduct of their affairs. Information was currency in these farming networks and needed to be guarded at all costs. You did not on any account disclose your business, let alone allow strangers to pore over your family records.

The second section brings us back to Norfolk and the development of a system of farming with partners and contract farming in the 1960s. This information relies heavily on the experience of two brothers who expanded their family business after the war using forms of sharecropping. The incentive was the growth of the frozen vegetable industry. With only a limited acreage, they increased their operation by 'taking' or 'borrowing' land for a season in return for a share of the output. The relationships formed with farmers and landowners led them to create a management and consultancy practice, whereby the brothers were entrusted with the farming in return for an agreement involving the sale of the crop. As with the family farmers of Shropshire, the historian's access to this enterprise was obtained through local connections and was based principally on oral evidence.

The final section considers the circumstances which led the Country Landowners Association to promote sharefarming amongst its members in the 1980s.[3] As we have seen, this initiative stemmed from the terms of the Agricultural Holdings Act, 1976, which compromised the property rights of landowners by extending tenants' security of tenure for three generations.[4] New methods of taxation, whereby rents were assessed as unearned income, also penalized landowners severely, making them liable to the full force of Capital Gains and Capital Transfer Tax. The aim of the CLA was to find a way round this punitive legislation. Their recommendations were based on practice in New Zealand, there being no institutional precedent for the system in this country. Insights were obtained from a range of landowners, including the late Ian MacNicol of Hunworth and Stody, lands once owned by Edmund Britiffe who drew up the agreements at Felbrigg in the 1670s.

The diaries of the family farmers from Shropshire remind us that one of the earliest recorded references to farming at halves was found in Robert Loder's farm accounts, a man of similar status farming in Berkshire 400 years earlier.[5] The diaries show men and women striving to take advantage of new opportunities in the years of economic recovery in the 1900s and during the Great War, and then struggling to make ends meet in the depressed conditions of the 1920s and 1930s. In each situation they placed livestock at halves, just as their forefathers had done centuries before. The most detailed evidence comes from the diary kept by William John Cooke from 1879 to 1931. He was the tenant of three farms in the parish of Ratlinghope, which lies on the western slopes of the Long Mynd, a few miles north east of Bishop's Castle. The farms comprised Coppice and Ratchup of 283 acres, and Upper Stitt of 306 acres which William Cooke leased from Mrs Scott of Ratlinghope Manor, and Lower Stitt of 45 acres. He purchased the first two farms when the Scott's estate was sold in 1920. David, the present owner of Coppice Farm, is descended from William's youngest son Wilfred. William married in the 1880s and continued farming until his death in 1931, when the diary ends.[6] It is not a formal diary, but a hard back account book with dates inserted by Cooke. The entries show two distinct hands, the second appearing in the early 1890s. This round and legible hand, possibly of William's young wife, brings order to the diary; random entries of domestic and farm payments are replaced with titled pages dealing with particular enterprises and accounts. From that time, it becomes principally a record of stock movements, grazing agreements, breeding details of cows and mares, and wages paid to farm servants and day labour, all of which involved specific dates and deals, and required careful noting.

The diary tells us about the farming year on a Shropshire hill farm. Most striking are the agreements to secure winter grazing for sheep, cattle and ponies from neighbouring farmers and landowners. The most common agreements involved paying a rate per animal per week, known as 'tack'. Under this regime, wethers, lambs, bullocks, cows, heifers, ponies and foals would be sent off the farm for a period usually extending from October to April.[7] In the spring, the livestock were 'brot back' and sent 'to the hill' – the Long Mynd, where the Cookes, like other farmers in the area, enjoyed extensive grazing rights. Before proceeding to the summer pasture, the heifers and mares were covered, and the sheep shorn, dipped and pitched. The arrangement for the ewes at halves was different. They would leave the farm in September or October for about 12 months, the grazier having taken responsibility for lambing, clipping, dipping and so on. In return he would keep half the lambs and half the wool. From 1917, all the wool was retained. The earliest of Cooke's agreements date from 1901. The later example refers to the 'usual conditions' indicating the well established and common nature of the practice.

> 1901
> Oct 16   Took 40 three year old Ewes and 1 Ram to Mr Lloyd, Farley on halves Viz. half the lambs, half the wool and half all other expenses the Ewes to be returned in good marketable condition at the end, not later than the end of August 1902
> Oct 19   Sent 15 Ewes to Mrs Garside, Ellesmere, to halves on the usual conditions, viz. to remain 12 months $^1/_2$ the wool half the lambs half all expenses, the ewes to be returned at the end of twelve months in good saleable condition …

There are eight more such agreements, the last one dating from 1925. It is not clear whether William Cooke's successor continued the practice, but possibly he did, judging by the regular appearance of advertisements in the local press. These agreements are similar in format and purpose to those dating from the seventeenth century. The grazing rights on the waste allowed farmers to hold animals well beyond the capacity of the home farm, but it required them to make arrangements for winter keep. Rather than bring heavily pregnant ewes back to the farm in the spring, the grazier undertook the lambing and clipping, and in return, as we have seen, kept half the lambs and wool. In this way the system provided inducements to the grazier, and allowed the farmer to make a tidy profit at the end of the year. Note that he wanted them returned in 'good marketable condition'. David Cooke says that the ewes were purchased as yearlings at the autumn livestock sales, and sold the following year as two year olds. The mutual benefit to both owner and grazier of the sheep leads us to wonder why there were not more agreements of this nature. But we know that they relied on trust between the parties which was not always forthcoming. Graziers were well known for having 'little sales' before the ewes were returned with 'half' the lambs; by the late 1930s farmers rarely bothered about the wool.[8] Nevertheless, the practice provided the farmer with an opportunity to deal and expand his business, particularly if he could find a regular partner, which William Cooke did by joining with W.J. Wood of Garmston from 1908. In the depressed conditions of the 1920s and 1930s, established farmers, like William Cooke, would restock bare hill farms with ewes at halves, helping small farmers back on their feet. Taking ewes at halves was known as the last stop before the workhouse; significantly there was no term used to describe the beneficiary.[9]

The other two farm diaries come from the lowlands around the market town of Cleobury Mortimer on the Shropshire and Worcestershire border. On these farms there was less urgency to find winter grazing; nevertheless, ewes at halves and sheep and cattle at tack, were used as a way of intensifying farming regimes and 'growing the business' without the need for buying or leasing extra land, and adding to fixed costs. By taking animals off the farm, it enabled farmers to grow more corn and make more hay to feed milking cows and bullocks over the winter. In his diary Richard Henry Evans shows how this strategy worked at Curdale Farm between 1908 and 1940. He leased the farm of 228 acres from the Longleat Estate, and like William Cooke, took the opportunity to purchase it as a sitting tenant in 1919. His wife Ellen Crump also owned a small farm at Bayton which they let. So they were 'well placed'. His returns in 1918 show that he cultivated at Curdale 33 acres of wheat, 9 acres of barley, 15 acres of oats, 2 acres of potatoes, made hay from 17 acres, ploughed 21 acres, and kept 24 milking cows, with replacements, bullocks, sheep and poultry on grass. Taking advantage of their proximity to Cleobury Mortimer, Ellen ran a milk and egg round, and they sold chickens, lambs and bullocks to the local butchers.

The first reference to ewes at halves appears in 1908, '50 ewes gone to Worcester at halves' (Type A). More common were arrangements with neighbours, as in 1909, '8 ewes at halfs at Tom Lamberts went to the ram on 18th Oct 1910'. Occasionally, cattle were involved, as in 1928 when '6 cattle to go to Mr Stones valued £10 per head at halves, profits when sold to be shared' (Type B). The small numbers involved and the variety of contracts suggest that Evans offered a valuable local service, providing credit and stock to small farmers and helping out tradesmen, widows and professionals with a few acres of grass or park land. In this way, Evans secured extra grazing for his stock and they found a farmer with the ability and expertise to manage their grass. It was an ideal arrangement for both parties with little money changing hands. By the late 1920s, Evans had established a regular partnership with Mr Stones and Mr Neath. In 1927, Neath dishonoured a cheque, which helps to explain why established farmers, like Evans, were not averse to payment in kind in this difficult period. By 1933, he had devised accounts and records to keep track of these transactions; as in the case of the Cookes, it may well be that this was the purpose of the diary. The last reference to ewes at halves dates from 1940, which coincides with the onset of war and the final advertisements in the *Shrewsbury Chronicle*. With government subsidies and the encouragement to grow corn and sugar beet, farmers did not have to bother with these sorts of petty arrangements. The entries in the diary continue to record the grazing of stock on the farm, and prices of stock sold at auction, but all mention of farming at halves disappeared.

The diary from Lower Elcott has only limited references to ewes at halves, but in this case the diary only exists for a single year, 1914, with just scrappy notes up until 1917. Benjamin Griffiths was less fortunate than his neighbour, Richard Evans in that he shared the tenancy of Elcott with his mother and two brothers, John and William. He made his way by specializing in stock breeding and letting out the services of his stud bull at 5s a cow. His photograph appears in the *Wellington Journal and Shrewsbury News* in 1931, alongside the agricultural luminaries of the day, as 'thrice the cup winner with highest aggregate points'.[10] The diary contains the summary of an account with Walter Amies of Tenbury Wells, showing him placing ewes at halves between 1914 and 1917. Like Evans, Benjamin, as sitting tenant, took the opportunity to purchase his farm in 1921. Eventually he managed to establish four sons on farms in the district.

The descendants of these resourceful farmers understand the working and motives behind farming at halves. John Griffiths of Overwood Farm remembers his father Charles, the eldest son of Benjamin, placing ewes at halves with friends nearby. Not very successful at farming, they had difficulty finding the money to buy stock, so his father supplied them with ewes for two or three years in the late 1940s. They kept half the lambs and all the wool. It was regarded as a form of credit for a family member, who could be entrusted with lambing and looking after the ewes. Similarly, Edward Foster, tenant of two large farms on the Apley Estate at Bridgnorth, remembers his father describing how the family had helped out a fellow tenant in the 1930s, by buying up his cattle, grazing them on the Severn meadows leased by the tenant, and agreeing to split the 'uplift' on the sale of the beasts (Type B).[11] This is a similar practice to 'half crease', or half the increase, a common activity in the Welsh Borders since medieval times, and on Cornish farms until the 1970s.[12] In both these cases, we were asked not to disclose the name of the tenant or the farm, an indication of the private nature of many of these agreements.

In a more commercial way, John Haywood of Wall Town Farm, Neen Savage, remembers his father placing ewes at halves in the 1940s and 1950s. The Haywood family have few written records, but John's mother, aged 91, recalled several agreements with graziers including sending 500 ewes to a Mr Simpson at Broadway, in the Cotswolds, in 1946. More locally, between 1950 and 1963, they sent 250 ewes to a neighbouring farm at Kinlet and 100 ewes to a farm at Rock, near Cleobury Mortimer. The arrangement was that George Haywood bought the ewes, and the grazier kept all the wool for his work. They shared the lambs in return for the feed for the ewes – grass, turnips, stubble and hay. John explained how his grandfather often provided credit and stock to small farmers in the district. Many of them could not read or write, so placing ewes at halves and animals at tack was a simple way of doing business with them. He also thought that the practice was associated with the need to move sheep regularly from pasture to pasture to avoid disease, working on the principle that a sheep's worst enemy is another sheep. Given the scale of the activity, evident in the newspapers, there may be some truth in this.

The diaries and recollections indicate that farming at halves was a well established practice in Shropshire in the first half of the twentieth century, while the newspaper advertisements confirm that it extended throughout the Welsh border counties. A limited sample of local papers included the *Hereford Times*, the *Shrewsbury Chronicle*, the *Montgomery Express and Radnorshire Times*, the *Oswestry and Border Counties Advertiser and Montgomery Mercury*, the *Bridgnorth Journal*, and the *Ludlow Advertiser*. Only the last two papers did not carry advertisements, although we know from the Cleobury Mortimer diaries and the oral evidence that the practice existed in those districts. The *Hereford Times* carried the most advertisements over the longest period from 1920 to 1959, and covered the widest area, stretching from Shropshire to Monmouthshire, Gloucestershire and into Wales. No other newspaper carried advertisements before 1925, or after 1939. Those for ewes to halves appeared in the papers from late July to November – mostly from the end of August to the end of October. Where the terms are specified, invariably the arrangement was 'half the lambs and all wool given'. A notice in the Oswestry paper in 1939 offered to take just a third of the lambs, and by this time several offered terms 'better than halves'. Frequently, a point was made about confidentiality, and enquiries were directed through a box number, confirming the element of stigma attached to the arrangement. As late as 1939, in the *Hereford Times*, 25 different farmers were advertising for men to take 'ewes at halves'. One of the latest advertisements in 1959 shows how even in a period of relative prosperity older farmers might still favour this type of arrangement. 'Farmer retiring, having flock of black faced ewes, would like to contact farmer that would take them on profit-sharing terms. Strictly confidential. Good opportunity for the conscientious ...'.[13]

This type of sharefarming was not confined to the Welsh Borders. In Derbyshire, in the early 1920s, setting out on his farming career, George Henderson found a situation with an elderly sheep farmer who offered to take him into partnership without capital – George having only £60.[14] Realizing that George 'would not stop long in any paid position', he offered him half the flock, keeping the other half to support himself and his wife. It was in effect an old-fashioned maintenance agreement. Henderson noted that these farmers kept no records and no valuations for fear there would be fewer sheep next time they counted. The only record was the bank book. Like the upland farms in Shropshire, these open hill farms were valueless without stock and labour. George rejected the offer on the grounds that the prospects were very limited as the market for lamb declined and the hills deteriorated. Similarly, in Devon in the 1950s, Robin Stanes remembers a local farmer letting him have 30 Dartmoor Greyface ewes at 'half crease', giving him all the wool and half the lambs, offering advice and providing him with much needed capital.[15] Like George Haywood at Neen Savage, the farmer, Arthur Ellis of Pittaford, preferred to have his sheep off the farm, his cattle at home, and make hay from the surplus grass. This prevented his land getting 'sheep sick' and allowed him to 'make a penny or two'.[16] Stanes also remembers another farmer with two large dairy herds. He did not pay the cowman a wage but gave him a percentage of the milk sales, and provided the cows, fields, buildings and extra feed. This is similar to the system of sharemilking in New Zealand, where the milker receives a third of the milk cheque. In Norfolk, James Keith, a migrant from Aberdeenshire before the First World War, introduced incentive payments on his run-down west Norfolk estate. Horrified by the loss of ewes, which he attributed to lack of care, he allowed the shepherds, 'all the lambs over 30 the score', or any over 30 lambs born to a score of ewes, which had a remarkable effect on the survival of ewes and lambs. When he started a pig unit at Waterden Farm, he paid the young Danish manager, £2 10s a week 'plus half the profit'.[17] Impressed by his performance he later granted him a tenancy. These examples all show established farmers using forms of sharefarming to encourage able young men to make their way up the farming ladder.

The evidence cited here is clearly only the tip of an iceberg. With more research the extent of the practice in the Welsh Borders could possibly be mapped and measured. However, given the nature of the evidence, the value of such an exercise might be limited, as only a fraction of the activity ever reached the newspapers. For the purposes of this book, it is enough to know that it was a well established feature in most local papers during the interwar years. It may be that in the future, a technique can be devised for reaching this material, but from the foregoing discussions, it is clear that farming at halves remained a covert activity amongst farming communities. Each of our farmers stressed the need for secrecy, and the stigma attached to such agreements. To accept ewes at halves was an admission that one had insufficient capital to stock ones own farm, and relied on the goodwill of neighbours. Another significant issue was the status of these animals. By law, since the Agricultural Holdings Act of 1883, landowners had no right of distraint over agisted livestock for the payment of rent.[18] This explains why modern tenancy agreements often contain a restriction against agistment, or the hiring of livestock for grazing, as landowners would be keen to avoid tenants lacking the working capital necessary to stock the holding. The three Shropshire farmers were tenants when they started placing animals at halves with their neighbours, so they might well have been contravening their tenancy agreements, which all parties would certainly want to keep private. In reality, we will never know the full extent of farming at halves.

The question we can ask is why farming at halves emerged as a common practice at this time. Was it simply that farmers, with improved education and greater awareness, kept better records, and that newspapers increased the range of their advertising?[19] Or were there particular circumstances that led farmers to revive and develop this ancient practice? The system of putting out ewes at halves was not exclusively associated with the depressed years of the 1920s and 1930s. On all the farms, for which we have diaries, it first appears in the 1900s as English agriculture, particularly in the pastoral counties like Shropshire, staged a significant recovery.[20] In the earliest diary, which dates from 1879, there is clearly a change of outlook in the mid-1890s as off-farm grazing agreements become more frequent and detailed. This coincided with the growing demand for dairy products and meat from a rising urban population, and the development of the railway network which enabled those products to reach the market in good condition. The construction of railway lines to remote border towns, like Craven Arms and Bishop's Castle, which held the livestock sales, and the later introduction of lorries for the collection of milk, greatly increased the opportunities for these hill farmers. From that time, it became worthwhile for them to increase stocking rates, intensify farming regimes and develop new enterprises. Better communications also enabled them to deal in livestock, and take advantage of small farmers with insufficient capital to buy at auction. As we have seen, expansion could be achieved by off-farm grazing, which allowed farmers to expand their business without buying or leasing extra land, and adding to their fixed costs. Farming at halves was a useful way of making money with minimal capital outlay, which explains why farmers continued the practice during the prosperous years of the First World War.

After the war, many tenants received notices to quit from their landlords.[21] Faced with increased taxation, death duties, 'low rents, the high cost of repairs and heavy tithe payments', landowners found it 'impossible to continue' and took advantage of high land values to sell their estates. If they had raised rents, the increment would have been liable for income and super tax, whereas none was payable on the capital proceeds of the sale of farms. They got out, 'while the going was good and the market land hungry'.[22] From 1919, sales of estates fill the pages of the *Shrewsbury Chronicle*, as hundreds of farms, dairy herds and sheep flocks came on the market. Benefiting from a few prosperous years, tenants bought their own farms, as did the Cookes, Evans, Griffiths and Haywoods. But the optimism of the new owner-occupiers proved short-lived, as the government withdrew support in 1921 with the repeal of the Corn Production Act. Prices for corn and livestock plummeted as cheap imports returned to flood the English market. In this climate, without the support of state, or landlords, and having to pay costly mortgages contracted when prices were high, these farmers were particularly hard hit. Not surprisingly they turned to methods, like ewes at halves, which had proved so cost-effective in the past. As the economic situation deteriorated still further, they placed advertisements in the local press, offering ewes to men needing to buy stock or wanting to utilize grass land without the bother of buying and selling sheep. The practice shows once again how forms of sharefarming could be used to eke out a profit and to mitigate risk in the most hostile economic environment. Its continuation after the Second World War, and into the 1950s and early 1960s, points to a system long embedded in the local culture. However, the generation that took over these farms, brought up on subsidies and guaranteed prices, could afford to forego such painstaking methods.

Viewed from a longer perspective, we can see that farming to halves re-emerged in a period of radical change in farming and rural society. The years before the First World War provided new opportunities for farmers, and saw landowners, increasingly penalized by taxation and poor returns, gradually withdraw from their role of supporting tenants and underpinning English agriculture. Successive legislation, promoting the interests of tenants at the expense of landowners also provided an incentive to sell.[23] With the sale of estates, the number of owner-occupied farms increased from 12 to 40 per cent between 1908 and 1950.[24] By 1927 the extent of owner-occupation was greater than it had been probably since the seventeenth century. Table 8.1 shows the steady decline of tenanted land over the last century.[25]

The trend from tenanted to owner-occupied farms appears to be significant. Not constrained by leases or the dictates of landlords, owner-occupiers were much more flexible and could pursue different avenues, enter into informal contracts and new enterprises and adopt modern technology. As the old tripartite class structure broke down, the culture in the countryside changed to one that was more amenable to collaboration and co-operation. There was a time between the two world wars, when new owner-occupiers had led a precarious existence, devoid of landlord capital, crippled by high mortgages and low prices

Table 8.1   Agricultural land tenure in Great Britain, 1908–2005

Year	Rented % area	Mainly rented % holdings	Owner % area	Mainly owned % holdings
1908	88	88	12	12
1922	82	86	18	14
1950	62	60	38	40
1960	51	46	49	54
1970	45	42	55	58
1980	42	34	58	62
1989	37	26	63	74
2005	26		75	

Source: M. Winter, 'Farm Tenure and property rights', (unpublished working paper, 1992); W. Gemmill, 'The Role of the new entrant', *Journal of the Royal Agricultural Society of England*, 166 (2005).

and with only the sketchiest support from the government. During this period, farmers were more or less on their own, genuinely bearing the total risk of farming. Many did not survive.[26] In these circumstances, the most enterprising resorted to self-help, sharing risk and supporting their less fortunate neighbours, using techniques long familiar to their forefathers. After the late 1930s, as the state intervened and assumed the role of underpinning English agriculture, self-help systems like farming at halves petered out. But as we shall see in other circumstances, it once again played a part in helping farmers to develop new enterprises and take advantage of new opportunities.

The evidence here is based on the experience of two highly enterprising Norfolk farmers, Jim and David Papworth, whose family have farmed in the Aylsham area since the 1920s. Their farms at Felmingham and Tuttington lie on the fertile loams, south east of Felbrigg, between the Blickling and Westwick estates, where we found farming to halves in the seventeenth century. Indeed, the Windhams owned Tuttington Hall, where David lives, and a substantial holding at Dilham a few miles away, which they also farm. Today, the Papworths' farming concerns extend across the Broadland region and into south Norfolk, the area so warmly praised by William Marshall in the 1780s.[27] In 2002, they were awarded the 'Excellence in Practical Farming Award' of the Royal Agricultural Society and contributed a biographical sketch to the journal, entitled 'Diversity and Diversification'.[28] In this article, Jim traces the origins of the farming business, and outlines their versatile farming philosophy, which has always combined farming with food processing, retail outlets and working with partners.

The Papworths, like so many successful farming families in Norfolk today, migrated to the county during the depression of the 1920s and 1930s, and took advantage of rundown estates and cheap farms. The migrants, hard-nosed working farmers, progressed by introducing new profitable enterprises, including dairying, pigs, poultry, vegetable and fruit growing. 'Grandfather' Papworth, a miller from Elsworth in the Cambridgeshire fens, dealt in livestock 'of all descriptions, cattle, sheep and pigs'. In 1924, he bought a small farm at Felmingham and later secured the tenancy of Tuttington Hall. Jim remembers his father, Leonard as a 'wonderful husbandman', who always insisted on the highest standards, kept meticulous records and made huge profits on his pig-finishing sideline. This attention to detail reminds us of the Shropshire farmers keeping their diaries, but unfortunately those of Leonard Papworth do not survive. Like the Evanses of Curdale, the Papworths benefited from their proximity to a vibrant market town. In the inter-war years, they similarly delivered milk and eggs, and supplied local butchers with poultry, beef and lamb. Jim did not know if his father developed his business by placing animals at halves or at tack, but we can not rule it out.

Jim's biography shows quite clearly that he is not an average farmer. He started his farming career at school rearing Guinea pigs for vivisection, sending them down to Chelmsford by train in cane baskets. With the capital from this enterprise, he secured a loan from the bank to set up a poultry business when he left agricultural college at the age of 16.[29] He hired buildings and delivered his poultry all over Norfolk, selling 'point of lay pullets', and cockerels to farmers' wives for fattening on the 'shack', or cereal stubbles for the Christmas market. In the 1960s, new opportunities arose with the development of the frozen vegetable industry by Birdseye and Ross Foods. Jim and his brother did not own a vast acreage, so they increased their operation by growing 'all manner of crops on any land that we could take or borrow for a season'. Taking a crop is a tradition well known in the Fens, and in these very fertile parts of east Norfolk, whereby the farmer allows the taker to grow a catch crop over the summer months in return for a share of the output. For example, once the peas have been harvested in June, a crop of dwarf beans can be grown before the land is sown with winter corn in November. The profile of the Papworth's crops at the time included spring cabbage, broad beans, dwarf beans, cauliflowers, Brussels sprouts and vining peas, as well as the more usual cereals, sugar beet and potatoes. The cultivation of these different crops required expensive and specialist equipment far beyond the resources of individual farmers. With their farming neighbours, the Papworths formed a farming co-operative, Aylsham Growers, to share the cost and maintenance of machinery.[30] In this way, ancient traditions of sharecropping and pooling equipment have facilitated the creation of a more diversified agriculture in this part of Norfolk. Their approach is very different from the simple system of corn growing promoted by the eighteenth-century agricultural improvers and practised by tenant farmers on large estates, which from the 1880s, proved to be the undoing of Norfolk agriculture.[31] In the late twentieth century, the Papworths used forms of sharefarming to effect the necessary diversification from corn into vegetable crops and livestock.

The arrangements Jim formed with farmers and landowners led to the creation of their contract farming business, which Jim believes was the first of its kind in Norfolk. The brothers devised their own particular system which involves the sale of the crop. It varies from partner to partner, depending on their particular needs and circumstances, but basically it works like this. The Papworths, as the contractors, set out their terms and conditions, agreeing to undertake full responsibility for the farming of the holding; they provide all the labour, machinery, chemicals and seeds. The parties then reach a figure for the crops, at so much per acre, based on the productivity of the land, facilities and assets. This figure, less the bill from the contractors, constitutes the 'margin', which forms the landowner's profit. The transaction is made just before harvest, when the contractors present their bill and buy the crop from the owner; they then sell the crop by the tonne on the open market – this is their profit. At first glance, there appears to be no element of sharing in the arrangement. But look more closely. The figure for the crop is negotiated by both parties just before harvest, while it is still standing, so they know the size of the crop. In the seventeenth century, assessors also divided the crop at this stage. The contractor then buys the crop from the owner and takes his risk on the open market; this deals with the problem of disposing of the crop. The return for both parties is determined each year by the size of the crop. In this way, they share the risks of cultivation, and the terms can be adjusted for good years and bad years. Evidently, with one farmer, no sum is even agreed beforehand, the arrangement is simply a matter of 'see how you get on ... if things go well I know you will look after me.'

The system relies on trust to a degree which seems quite extraordinary to modern eyes. At no stage, apparently, do they involve agents, lawyers or mediators. The agreement has no legal basis, so transactions costs are non-existent. Each party retains the right to walk away from the agreement, with no recriminations. However, as it is in the interests of both parties to make the system work in this way, the partnerships, as they are called, survive in some cases for many years. The key to success is the reputation for trustworthiness and excellence in farming that the contractors have built up over the decades. This is their trademark. They now own land on a considerable scale, and farm it themselves alongside the land of their partners. The close personal relationships, their understanding of the locality and its economy, shows why their system works today, why these arrangements survive in tightly knit rural communities, and of course, why they remain hidden and secret; 'everything is confidential'.

The demographic profile of the Papworths' partners is reminiscent of the social context in which these arrangements have appeared in the documents from the earliest times. Of the 25–30 partners, the largest number in the present arrangements are elderly couples, sometimes childless, or with daughters or sons disinterested in the farm. Several are widows or spinsters, who simply want to end their days in their own home, but need someone, if possible a trusted friend, to take on the responsibility for farming. A few are existing tenants, who have diversified into business, and effectively sub-let their farming operation to the Papworths. The Papworths are also engaged by owners of landed estates and country houses, who wish to keep a measure of control over their premises, particularly parkland. Neighbours, with adjoining parcels of land, frequently hand over their few acres to be cultivated. Other farmers have lost interest, or are too infirm to farm, precisely the kind of reasons which prompted Robert Loder to 'put to halves' his mother's jointure in 1610, and John Berney to let to halves the holding of a widow at nearby Westwick in 1698. As it happens, the Papworths have themselves farmed Dilham Hall Farm, which was let to halves in the 1650s and 1660s, and Intwood Hall, where John Brewster implemented his scheme of contract farming in the 1670s.

The size of the holdings managed by the Papworths vary from a few to several hundred acres. Most are between 50 and 250 acres, but five are over 300, seven less than 50, and three less than ten acres. Every proposition is assessed on its merits, no acreage is too small, if the situation fits. In recent years, they have narrowed their cropping profile, as the market has moved from frozen to fresh vegetables, which are more difficult to supply from this part of Norfolk. As an alternative, they expanded their sugar beet contracts by forming another co-operative, the White Cross Sugar Beet Group, which allowed them to utilize the capacity of high-output six row sugar beet harvesters. Making maximum use of 'ever improved equipment', through co-operatives, remains a central part of their strategy. In 2003 they grew about 2,000 acres of sugar beet and supplied around 70,000 tonnes to British Sugar's factory at nearby Cantley.

Perhaps the most imaginative aspect of their business has been the development of their livestock enterprises. From the 1980s, several partnerships have involved the management of parkland and marshes, often important environmentally or for landscape reasons. Capitalizing on support grants available for such areas the Papworths built up a complex network of livestock rearing and fattening units. They channel the finished animals, pigs, lambs and cattle, through their own abattoir and two butchers' shops at Fakenham and Swaffham, completing the links between farming, food production, retail and consumption, or from 'farm to fork'. With their widespread activities, the Papworths have been able to dovetail the enterprises on light and heavy soils, and marshland, with success 'beyond our wildest dreams'.[32] Sharefarming has given them the flexibility necessary for such a diverse and fast-moving farming business. The Papworths are not interested in buying more land, 'we see an increase in acreage, but not in landowner-ship'; their preference is for farming with partners, and the flexibility it provides, not the acquisition of a huge landed estate.

The long-term relationships established with farmers and landowners has led to the creation of a formal management consultancy, Oasis Man-agement, 'utilizing our own experience ... and offering down-to-earth opinions to estate owners, farmers and professional bodies'. This is not dissimilar to the advisory service offered by the Britiffes in this part of Norfolk in the late seventeenth century. In fact, in their methods, policies and outlook, the Papworths closely resemble the Britiffes, taking on dif-ficult leases and holdings, dealing in corn and stock, drawing up manage-ment plans and letting-to-halves agreements, forming local networks and selling their practical expertise to leading county families. Their experience, like the Britiffes', suggests a tradition of unconventional agreements in this highly fertile region, which like the Fens permits diversity and diversification, into a wide range of crops and enterprises. In both cases, these agreements have come to the surface in periods when the landlord-tenant system was not so dominant, in the late seven-teenth and late twentieth century. This may be due to the chance survival of written records, but it does indicate the greater flexibility associated with owner-occupation and unconventional agreements outside the classic landlord-tenant system.

The reasons for the revival of sharefarming in Shropshire in the first half of the twentieth century, and its development by enterprising Nor-folk farmers in the second half of the century are not difficult to under-stand. For these tenants, and later owner-occupiers, it has provided a low cost strategy for expanding their farming businesses, and allowed them to retain maximum flexibility for diversifying into new and more profitable enterprises without incurring huge fixed costs. It worked equally well in the depressed conditions of the 1920s and 1930s, and in the post war period of opportunity, when English agriculture was fully supported by a system of government subsidies and guaranteed prices. In enterprising farming circles, it came to be recognized as an effective way of sharing risk and avoiding the taxman.

The interest in sharefarming shown by landowners in the 1980s has different origins. As we have seen it was specifically linked to the passing of the Agricultural Holdings Act, 1976, which severely encroached on the property rights of landowners. This law extended tenants' security of tenure from one lifetime to three generations. It was the continuation of policies introduced by radical governments from the 1880s, which had reinforced the position of tenants at the expense of landowners.[33] At the same time, the government introduced new taxation which penalized landowners, rents being assessed as unearned income and estates liable to the full force of Capital Gains and Capital Transfer Tax. These measures deterred landowners from creating tenancies and forced them to seek other ways to farm their land. Increasingly, when landowners had the opportunity they kept farms in hand, appointed managers or employed outside contractors. However, these methods necessitated a degree of involvement which did not suit all landowners, nor were they popular with local communities who like to see farm houses occupied on a long-term basis with established tenants contributing to the life of the village, supporting the local school, church and shop. Partnerships offered another solution, but beyond the family few landowners would countenance the risks involved. It was in this context that the Country Landowners Association commissioned a survey of sharefarming practice in New Zealand to find out whether it could be adapted to British conditions. The report by Richard Stratton became the bible for modern practitioners.[34]

The virtue of a sharefarming agreement, and the feature which distinguished it from a partnership, or a simple contract, was that the owner was not liable for the debts of the operating farmer. Owner and farmer share the inputs and share the output, in agreed proportions, but essentially they conduct separate businesses. In theory, sharefarming met all the requirements of landowners: it created no tenancy, landowners retained flexibility and control over their land, and the element of participation and risk meant that the landowner's share in the agreement became eligible for assessment as earned income. Despite the obvious advantages, most landowners remained sceptical; the idea of *sharing*, and its association with sharecropping and colonial economies, seemed totally alien to British tradition. They wanted their own system reformed, and campaigned vigorously for the Farm Business Tenancy (FBT), which was secured in 1995, ending security of tenure, and placing the relationship between landlord and tenant on to a fully commercial footing. It was heralded as a completely new relationship between landlord and tenant, but to the historian it looks like a straightforward eighteenth-century lease.[35]

In the intervening years, sharefarming became an established practice amongst a significant minority of landowners, several of them leading lights in the CLA. The National Trust also implemented a number of schemes, notably at Hardwick Hall, Derbyshire, and Charlecote Park, Warwickshire. Land agency firms, particularly those associated with Richard Stratton's practice in Truro, Cornwall, developed an expertise in setting up these arrangements, and pockets of activity appeared across the country. During this period, the Royal Institution of Chartered Surveyors, commissioned an academic enquiry into the state of agricultural tenure in England and Wales. The research, led by Michael Winter, uncovered a much higher rate of unconventional tenure than anyone had predicted, 'a rare example of applied research findings giving genuine surprise to the client community'.[36] As we shall see, it found that a significant proportion of agricultural tenancies were held on unconventional agreements and included an element of shared ownership, and that this figure was most likely grossly underestimated.[37]

With the advent of the Farm Business Tenancy in 1995, many predicted the demise of sharefarming, but this did not occur. In fact, land agency firms record a marked growth of interest in 'joint ventures' in the 1990s and 2000s, as farm incomes continued to decline. Once again, sharefarming, or unconventional tenures, offer the optimum solution in times of difficulty. What follows is a sample of the experience of land agents and landowners, the results of a limited enquiry conducted in 1998 and 2002 before the practice became universal. The agents were asked to respond to a questionnaire covering such issues as the benefits and disadvantages of the system, the motives for entering the agreements, the circumstances in which it occurred, and whether sharefarming offered a way forward for British agriculture. Overwhelmingly, the flexibility and the measure of control it allowed, and the involvement of the landowner was considered a major benefit. The tax advantages were an important consideration, but were rarely the principal reason for enter-ing an agreement. More significant was the perception, offered by Peter Fletcher, of Stratton and Holborow, that it was a 'fairer system' all round, 'when returns are good, both parties benefit. When returns are poor both parties share the agony!' This was not possible in a tenancy situation where the rent was fixed for three years, 'Agriculture moves much more quickly these days'. The idea of cultural, or socially driven motivation caused a stir amongst the respondents, leading Fletcher to refine his position, more as enlightened self interest, 'I would say that the landowner is to an extent still driven by economic reasons and the hope, of course, would be that there are more good times than bad, so overall and taken over a period of say ten years, the landowner would benefit more by sharefarming than he would through a straightforward letting'.[38] The main disadvantage was the risk of diminishing returns in hard times, the need to input capital and the 'hands on' involvement. John Young, a former agricultural adviser to the National Trust, also considered the procedures burdensome and expensive, but this, not surprisingly, was strenuously denied by the agents. The vital ingredient was the parties themselves. As we saw in the late seventeenth century, trust, flexibility and the willingness to co-operate was the key to any successful share-farming agreement. Agents were generally more sceptical about whether it would invigorate rural life by offering incentives to young entrants, as the costs of entering farming were simply too high for youngsters without capital. Sharefarming was considered an important facet of land use, but not the sole answer for British agriculture; it relied too much on particular circumstances and the aptitude of individuals. John Young was convinced, in 1998, that sharefarming would be overtaken by the Farm Business Tenancies. But the agents remained optimistic that their share-farming agreements would survive.[39]

FBTs have not, in fact, proved universally popular. In such uncertain times, tenants resent the loss of security and appreciate the involvement and support of the owner. The most recent responses from agents confirm the acceleration of co-operative and joint ventures, formal and informal. According to Philip Wynn, of Aubourn Farming Ltd, a subsidiary of FPD Savills, the greatest speed of change is in the east [of England], as arable units are more easily 'put together' than those for livestock. Most agreements were based on combinable crops and sugar beet. Intensive cropping and vegetables were not normally seen in these ventures as landowners tend to let them for intensive cropping for the season, rather than entering into a long-term arrangement – similar to the Papworths' arrangements. Dairy agreements are becoming more commonplace, but are much more complex to initiate and run because of the nature of the business. They have to account for live-stock numbers and values, and a quota, which can be variable. These different circumstances remind us of the difficulties encountered in the late seventeenth century at Felbrigg when leasing dairies was com-bined with letting to halves agreements, and the success at Raynham, when letting to halves was restricted to corn production. The modern arrangements resemble the nineteenth-century profit-sharing schemes in that both parties take 'a first charge' and share the profits, rather than simply divide the crop. But the contract is supervised by the agent, not the whim of the landowner. Wynn is in no doubt that with the severe pressure on farm incomes, all types of collaboration will con-tinue to be taken up in order to reduce the costs of production and take advantage of further economies of scale. Although new entrants to farming have historically focused on these types of arrangement, they are now commonplace across the sector.

Strutt and Parker have been associated with contract farming and sharefarming for over 20 years and provide a sample pack clarifying the principles and showing how it works in practice. The agreements set out the shared role of farmer and contractor, specifying the provision of capital made by the farmer: land, buildings and fixed equipment and breeding livestock, and by the contractor: labour, machinery, management expertise, and livestock if not provided by the farmer. Their particular functions are identified, and the responsibility for variable and fixed costs form the first charge on the account. The financial arrangement is that the total output is calculated, from which all costs applicable to the agreement are deducted. These fixed costs include the 'Contractor's Basic Fee' and the 'Farmer's Prior Charge'. What remains is 'The Divisible Surplus', the ratio being fixed in the contractor's favour, but not always. The attached Dairy Farm Contract Arrangement shows a 50:50 division.[40]

The growth of sharefarming over the last few years can be explained by the deteriorating economic climate, encouraging farmers to look for the most flexible and tax efficient systems. But there can be little doubt that the spread and accessibility of information has played a major part in developing awareness at every level. Joint ventures are promoted online, in agricultural colleges, recommended by DEFRA and by a host of bodies as a way of advancing environmentally friendly and sustainable agriculture. At the same time, as more and more farms are sold to businessmen and those in search of the 'good life', sharefarming is seen as a most acceptable way of cultivating the land. It offers the new enthusiast the opportunity to participate in farming – without the responsibility – and to forge links with the local community. From the farming perspective, the infusion of new capital – and new people – is most welcome. In other words, sharefarming goes some way to meeting the economic needs of the agricultural sector, makes a valuable contribution to the social structure of rural villages and suits the inclusive culture of caring for the countryside, which prevails today. With the current high prices for grain, there is yet another reason for sharefarming, as landowners, like their medieval forbears, can share in the bonanza.

From the landowners' perspective, a desire for involvement in the countryside motivated some of the earliest practitioners of sharefarming. Notable is the experience of John Henderson, who has worked his 1,750 acre estate in West Yorkshire with two sharefarmers since 1984.[41] He was greatly influenced by Richard Stratton. Indeed, Stratton's booklet on the merits of sharefarming came out under his chairmanship of the CLA's Agriculture and Land Use Committee, and he actively promoted sharefarming amongst farmers' groups in the early 1980s. Significantly, he was a woollen manufacturer and enjoyed the challenge and risk of sharefarming, particularly the pooling of ideas and the creation of a business environment in the countryside. Similar sentiments were expressed by John Cyster who in the 1990s farmed at Newenden, on the Kent/Sussex border, and supplied the London market with luxury dairy products.[42] Like-minded landowners were keen to find partners with entrepreneurial flair, willing to take a risk with new products, create 'niche markets' and employ local labour. They emphasise the view that there is a real need to bridge the gap between those with capital and those with the expertise and ideas. The problem is not the availability of youngsters, but the willingness of established farmers to break away from traditional arrangements which block the path of new enterprising entrants into farming. A greater awareness of sharefarming and its virtues is essential if these barriers are to be broken down.

But what of Norfolk landowners? How have they responded to modern forms of sharefarming? The interviewee in this case was the late Ian MacNicol, a former president of the CLA and owner of the Hunworth and Stody Estate.[43] The estate had been built up by the Britiffe family in the seventeenth century and passed to the Hobarts on marriage in 1717. Amounting to 3,000 acres it formed part of the Blickling Estate until it was sold to pay death duties in the 1930s. Ian MacNicol adopted forms of contract or sharefarming from the early 1980s. Like John Henderson, whom he knew well, he was influenced by Richard Stratton and promoted the practice amongst farming groups in Norfolk. But his approach remained flexible with each arrangement tailored to meet the requirements of the particular situation. Sometimes he undertook the role of contractor, like the Papworths, and on other occasions he offered sharefarming agreements to tenants. His first experience was with a farmer, whose lands lay close to one of his offlying farms. He agreed to farm the holding with his own, effecting significant economies of scale to the benefit of both men, while the former pursued a separate career selling agricultural products and services. They engaged a lawyer to oversee the agreement, but it was essentially informal and relied on the trust between the parties. They shared the decisions – depending on their expertise – and the provision of capital. As contractor, MacNicol started on a 'Contracting Fee' and charged for his use of labour and machinery. The former received the 'Prior Charge' or rental equivalent for his land and buildings. All other expenses were charged against 'the

Farm Account'. At the end of the year, the 'Divisible Surplus' was calculated, at a ratio of 80 per cent to the contractor. This is the classic arrangement used on the estate and is modified as required.

On another occasion, in the early 1980s, a tenant died leaving two daughters. MacNicol wanted to avoid the 'inherited' tenancy passing to a son-in-law, but did not wish to deprive him of his livelihood, so he offered him a sharefarming agreement. It was the typical arrangement, the landowner providing the land, the tenant the labour. But in this case, MacNicol also purchased the stock which the tenant fed and cared for and they divided the profit. Although he favoured FBTs, MacNicol continued with contract farming from 1995. In some cases he had no option. He tried to persuade a retired farmer, who asked him to farm his land, to grant him a FBT, but was refused. The farmer wished to remain 'involved', attend meetings of local farming groups, and, of course, retain the tax concessions available to a working farmer. Also, in his social circle, it was more attractive to have a contract and a 'deal' with Mr MacNicol of Stody Lodge, than sever his links with the industry. As we saw with the Papworths' partners, this motive is very common amongst elderly farmers reluctant to relinquish their connection with farming. In an agreement extant in 2003, MacNicol paid by the hectare: sugar beet £90, cereals and peas £80, set aside £12. After this first charge, it followed the usual arrangement of a divisible surplus with a 80–20 per cent split in favour of the contractor and above a certain figure the division of 50:50. Also in the contract, 40–50 acres of potatoes were farmed by a third party, reflecting the different specialist machinery required for the crop. A major advantage of contract farming is that it allows farmers to specialize and spread the costs. All things being equal, MacNicol would have phased out contract farming in favour of FBTs. In his experience, the transaction costs of setting up and administering the contract were expensive. Fees had to be paid to agents for negotiating with the parties and accountancy charges were higher than usual as it required the most detailed records. Moreover, with the FBT, the owner received a guaranteed return.

The complication with FBTs lay with the future of EU payments. In the Mid-Term Review of the reform of Common Agricultural Policy, it was proposed that payments ceased to be linked to productivity. On the basis of a three-year average, payments would be paid as a direct subsidy to the person farming the land to spend at his discretion. The controversial aspect was that tenants would have the right to take that 'entitlement' without the land, stripping the farm of its entitlement. To prevent this happening, landowners avoided creating tenancies, just as they did with the Agricultural Holdings Act. If they engaged a contractor or sharefarmer, the owner remained the working farmer, received the payment and safeguarded the entitlement. The likelihood was that the new arrangement would, as in the 1980s, lead to a further increase in contract and sharefarming. In his forecast, MacNicol proved to be right.

In their advice to farmers and landowners, in the aftermath of the Mid Term Review, Savills recommend contract agreements over the Farm Business Tenancy.[44] The virtue of the well structured contract involved 'sharing the risk between the parties with an annual management account to allocate the relevant reward for the respective risks'. The Single Payment rules had also resulted in the 'rebirth of the Profits of Pasturage agreement ... this archaic agreement involved the sale of the standing crop of grass without conferring any occupancy right on another party'.[45] This allowed the landowner or farmer to claim Single Payment on the grazing area. In these recommendations we recognize old and familiar methods. New entrants to farming, faced with high land prices and lack of working capital follow a similar course, 'Sharefarming arrangements, or some other joint venture of partnership, are still the most common way for a new entrant with limited resources to get into farming ... a number of very successful entrants have started this way, often by finding an older business partner who ultimately wishes to retire out of farming, but is looking for a go-ahead young entrepreneur to carry the business forward.'[46] At Hardwick Hall, the National Trust overcame earlier reservations and granted a new Share Farm Agreement in 2005 to the tenants for the management of the parkland with two rare breeds; 'it has given us the opportunity and confidence to take on a larger farm than we would have been able to under a tenancy'.[47]

Gaining access to a farm has always been a common reason for share-farming. The wheel has turned full circle. European legislation is accelerating a trend back towards sharefarming and unconventional agreements; they started in the early twentieth century and gained momentum again in the last quarter of the century. As the landlord-tenant system continues to decline, new entrants, farmers and landowners are reverting to methods long familiar to their forebears, before agriculture was so efficiently organized to maintain the interests and lifestyle of a ruling landed elite. When that political power faded it was only a matter of time before more flexible and cost effective methods of organizing agricultural production would be revived. As Strutt and Parker point out, sharefarming is particularly relevant for farmers wishing to restructure, reduce day-to-day involvement, improve cash flow, release working capital and maintain tax benefits. Today, sharefarming once again provides answers for land-owners, and continues to be a useful option for family farmers. In the next chapter we provide statistical data to flesh out this anecdotal and empirical evidence, and the conclusion will offer the opportunity to draw together the threads of this history of sharefarming.

9
Sharefarming at the Turn of the 21st Century

(co-authored with Michael Winter)

As its title suggests much of this book has been devoted to uncovering the hidden history of sharefarming. It has been hidden for a variety of reasons, not least because until the end of the twentieth century there has been no systematic attempt to record its incidence, despite the fact that the state has been collecting agricultural statistics since the nineteenth century.[1] This chapter discusses some of the limitations of government statistics and shows how they can effectively hide the presence of sharefarming. In order to attempt a more accurate quantitative assessment of the extent of sharefarming at the start of the twenty-first century, we present the results of a contemporary survey of the extent of sharefarming in 2007.

The use of statistics in the growing jurisdiction of the state, which reached a peak during the Second World War, rendered agriculture 'visible' in a new way.[2] However, statistics render visible only that which is measured. The statistical endeavours of twentieth-century bureaucrats reflect for agriculture, as for any other sector, both policy concerns and assumptions. In the case of land occupancy, the gathering of land tenure data went hand in hand with successive rounds of legislation that gradually increased tenants' rights and security, at the expense of landowner power. The extent of owner-occupation grew from 12 per cent to 62 per cent of agricultural land in Great Britain between 1908 and 1988.[3] During this period, the politics of land ownership centred on the relative economic merits of owned and rented land, and the moral and social questions surrounding the question of large estates and the continuation of the landlord-tenant system.[4] Not surprisingly, the prime statistical distinction has been between rented and owned land with scant recognition in official data of the complexities, and possibilities, within the categorization of 'owned' and 'rented'. This simple distinction can be seen in the annual agricultural census data and the occasional surveys, such as the wartime National Farm Survey.[5] Asked to complete a census form, indicating whether land was owned or rented, a farmer would do so, not in terms of an academically rigorous definition, but according to formal, or in some cases informal, contractual arrangements, often determined with the taxman in mind.

Two well-known examples illustrate how statistics derived in this manner distort the underlying realities of land occupancy. The first was noted by Newby and his colleagues in their seminal work in the 1970s on the sociology of agriculture in Suffolk.[6] They found a significant incidence of land being recorded as tenanted when it was owned by a family trust with one partner in the trust acting as 'tenant'. Thus the tenant had a share in the ownership and often a strong prospect of eventually having a fuller share or even sole ownership. The second example was, in a sense, the mirror opposite. A farmer, who no longer wished to farm land in-hand, for tax purposes might wish his income to be treated as earned agricultural income as opposed to unearned rental income. A course of action, commonly followed in these circumstances, was to 'sell' a standing crop, usually grass, which the 'buyer' had access to for a prescribed period of the year. Such land continued to be recorded as owner-occupied in official land tenure data.[7] These two examples indicate that, as in earlier centuries, it is possible that farming to halves is hidden even in an era of much greater statistical visibility. Certainly, it became apparent to many by the 1980s that the official data on agricultural land occupancy had become highly misleading in many respects.

The Agricultural Holdings Act of 1883 shifted the focus of farm tenurial arrangements from social relationships on particular estates to national statutory provision. Further statutes designed to protect tenants' rights to compensation and guaranteeing freedom of cropping followed, culminating in the Agricultural Holdings Act of 1923 regarded as the 'Magna Carta' of agriculture.[8] The 1923 Act consolidated the legislation and set down a full code for tenants' rights to compensation. Full security of tenure, however, did not come until the Agriculture Act of 1947 and the Agricultural Holdings Act of 1948, which provided life-time security of tenure to existing short-term tenancies of two years or more. Subsequently in 1976, the Agriculture (Miscellaneous Provisions) Act extended security to two successions. In the eyes of many agricultural commentators, what had begun at the end of the nineteenth century as a thoroughly laudable attempt to provide protection and rights to disadvantaged small tenant farmers, by 1976 had turned the tables to such an extent that landowners had virtually no incentive to let out any land that happened to fall vacant. Protection of tenants clearly benefited existing tenants hugely, but did nothing to facilitate new tenancies, the continuation of tenancies at the end of secure terms, or a farming ladder. The first piece of legislation that sought to redress the balance was the Agricultural Holdings Act of 1984, consolidated in the 1986 Act, which provided for the *opportunity* for fresh tenancies to be created for one life only. However, it did not rescind the rights of tenants granted under the 1976 Act and, given the retrospective nature of that legislation, most landowners were deeply sceptical about the guarantees offered under the 1984 Act in the event of a future Labour government.[9]

The inadequacy of the 1984 Act, combined with the reluctance of the Conservatives to tamper further with the legislation, led to much discussion about the extent to which farmers and landowners were seeking to circumvent the law by resorting to avoidance strategies, which included sharefarming schemes.[10] This speculation played into the hands of free-market Conservative thinking and the desire for a radical review of tenancy law. For some in the party, agricultural holdings legislation was an anathema, an undue interference in the market place that, amongst other things, restricted the flow of new entrepreneurial blood into this heavily subsidised sector.[11] It was almost as if the 1984 Act had been deliberately designed to fail, as a ploy to force the National Farmers' Union to consider freedom of contract, and to bring them closer to the views of the Country Landowners Association, Tenant Farmers' Association and the Royal Institution of Chartered Surveyors. The Conservatives were loath to tackle a problem without the prior agreement of the key interested parties, especially when two potentially opposed parties were both 'natural' party supporters, in this case the farmer and landowner organizations.

It took some time for the stalemate to be broken, and this occurred only when incontrovertible evidence showed that the law was increasingly out of step with land tenure realities. The Royal Institution of Chartered Surveyors, whose members stood to gain from either a tightening or a freeing up of the legislation, launched an academic inquiry into the state of agricultural tenure in England and Wales. The research uncovered a much higher rate of unconventional tenure than anyone had predicted; it was a rare example of applied research findings giving genuine surprise to the client community and consequently having a real impact on the policy debate.[12] The study of 1,790 farmers found that unconventional tenures – land not owned or rented under a Full Agricultural Tenancy – formed a highly significant element of farming in the late 1980s as Table 9.1 shows. One in five farmers in England

Table 9.1   Agricultural land area by tenure type (weighted data raised to national level) in England and Wales, 1989/90

Holding type	Hectares	%
Summary:		
Owner-occupied	6,959,057	58.7
Tenanted	4,891,291	41.3
Total	11,850,348	100.0
Formal conventional:		
Full Agricultural Tenancy with no share in ownership	3,204,484	27.0
Full Agricultural Tenancy with share in ownership	461,452	3.9
Total Full Agricultural Tenancy	3,665,936	30.9
Formal Unconventional:		
MAFF Approved Letting/Licence	69,427	0.6
Gladstone v Bower Agreement	100,225	0.8
Partnership	269,668	2.3
Share Farming	95,004	0.8
Total Formal Unconventional	534,324	4.5
Informal Unconventional		
Gentleman's or Informal agreement	209,324	1.8
Grass Keep	304,016	2.6
Cropping Licence	11,902	0.1
Sub-tenancy	24,743	0.2
Other	141,046	1.2
Total Informal Unconventional	691,031	5.9

Source: Butler and Winter, *Agricultural Land Tenure in England and Wales*, 2008.

and Wales occupied land on an unconventional arrangement, with grass keep and gentlemen's agreements being the most common. The findings demonstrated conclusively the extent to which the existing legislative provision failed to meet the changing requirements of the farming industry.

The RICS report provided firm evidence for the policy debate on tenure reform in the early 1990s. In 1991 the Government published a consultation document seeking proposals that would construct an enduring framework for the sector, deregulate and simplify tenancies, as well as encourage the letting of land. The result was the Agricultural Tenancies Act of 1995 which introduced Farm Business Tenancies (FBTs). In essence, the Act had three main aims: to encourage more letting of agricultural land; to increase the opportunities for new entrants; and to promote economic efficiency in agricultural land use.[13] To meet these aims, the provision of short and longer-term FBTs enabled a degree of flexibility in the land rental market and allowed parties a greater degree of freedom to negotiate agreements to suit their needs. FBTs were seen by some as offering the possibility of legal protection to parties involved in the more flexible arrangements which had come to characterize the informal sector. In 2006, an amendment to the 1995 Act signalled a desire to encourage diversification by tenant farmers, which would contribute to maintaining the viability of tenanted farms and improving still further flexibility in the tenanted sector.

Research undertaken in the aftermath of the emergence of FBTs did not follow up the issue of unconventional occupancy. It concluded that the 1995 Agricultural Tenancies Act had led to a significant amount of additional land being made available to let, but found it hard to disaggregate genuinely new land and that which had been converted from grazing licences, Gladstone v Bower lettings, contract and share-farming agreements. In short, research on FBTs told us very little about the extent to which this new more flexible form of tenure might have reduced the extent of any modern form of farming to halves. There is a persistent tendency – a frustrating one for those interested in the informal and unconventional – for research on agricultural tenure to focus exclusively on the complexities and challenges of the formal sector.[14] Fortunately, two other studies provide a more balanced account. A major study of 'joint venture farming' was undertaken for DEFRA by ADAS (2007) and a year later a re-study was undertaken, funded by the RICS, of the 1990 tenure survey.[15]

A fundamental problem in seeking to unravel the complexities of both unconventional tenure and farming to halves is one of definition. At heart lies a tension between the characterization of tenurial arrangements on the one hand and occupancy arrangements on the other. Often the characteristics of tenure and occupancy coincide, for example when a farmer both owns and occupies a piece of land or when one person owns and another rents land under a Full Agricultural Tenure. Occupancy and tenures can be described in identical terms in these instances. But our interest lies in a discrepancy between occupancy and tenure, when conventional language describing tenure is inadequate to capture fully the occupancy arrangements. For example ADAS defines joint venture farming as 'the bringing together of land, capital and skilled management in an agreement between two or more parties, each running their own business, rather than forming a new partnership'.[16] This definition is important, for the exclusion of formally constituted business partnerships means that the definition of joint venture farming might exclude some arrangements that we would characterize as sharefarming, particularly any partnership based on shared profits between a landowner and another farmer. As ADAS explain 'JVF farmers run their own businesses, they do not share overall profits (as in a partnership) but bear their own risks and derive profit from their own venture. However, they work with another business (or more) jointly to provide all the resources needed for farm production.'[17] ADAS offer a six-fold taxonomy of Joint Venture Farming:

Contract Farming – An operator (contractor) carries operations out on the land supplier's behalf for a fixed rate (£/ha) with a division of any surplus after costs specified in the agreement.

Contract Rearing – The operator/contractor (who can be an owner-occupier or tenant) rears animals for a livestock owner for a fixed fee per animal for a set time period or to an agreed specification (e.g. weight or heifer in calf).

Sharefarming – Farming operations are jointly managed by the land supplier and operator with some inputs and all outputs shared on a percentage basis, with land and buildings given a value.

Labour & Machinery Sharing – Two or more persons pool all or part of their labour and machinery and treat the arrangement as a cost centre or separate business which provides a labour and machinery service to each co-operating farming business for a fee.

Machinery Sharing – Two or more persons share in the ownership in all or part of their machinery treating the arrangement as a cost centre or separate business which provides a machinery provision service to each co-operating farming business for a fee.

Labour Sharing – Two or more persons pool all or part of their labour treating the arrangement as a cost centre or separate business providing a labour provision service to each co-operating farming business for a fee.

Although it might be argued that all of these arrangements derive in some measure from the sharing of risk, as well as returns, which denotes our understanding of farming to halves, it is clear that only two categories – contract farming and sharefarming – approximate closely to farming to halves in its historical context.[18] Unfortunately, the analysis undertaken by ADAS combines the two types of contract farming. Drawing on the annual Farm Business Survey (FBS) undertaken for DEFRA (using the 2004 sample of 1,789 farms), ADAS suggests that 1.17 per cent of farms were involved in sharefarming and

2.24 per cent in contract farming in 2004, a combined percentage of 3.41 per cent. According to ADAS this is a good estimate because the FBS is based on a randomized sample designed to be representative of all farms in England.[19] However, participation in the FBS is demanding of individual farmers. Although it is true that there are strict quotas based on farm size and enterprise type, a great deal of persuasion is required to encourage farmers to participate so as to meet the different quota targets. In particular, farmers are persuaded by the offer of benchmarking data, which allows comparison of their own business with others, to enter the survey. But the benchmarking data applies best to conventionally occupied farms and is less likely to be helpful to share and contract farmers who therefore have less incentive to enter the FBS. Thus, it is quite likely that the ADAS/FBS figures are an underestimation.

The 2007 survey undertaken for the RICS attempted a more accurate estimation. The key definitions of relevance to farming to halves are as follows:

> **Contract farming** – An agreement whereby the contractor carries out operations of husbandry as an agent for the landowner (or tenant). The landowner (or tenant) provides the land, buildings and fixed equipment, quotas (if applicable) and bank account. The contractor provides the labour, machinery and management expertise and is remunerated by an agreed formula.
>
> **Partnership farming with the landowner** – A partnership involving a farmer and a landowner in which the parties run the farm as a joint business.
>
> **Sharefarming** – A sharefarming agreement is an arrangement usually between two parties, a landowner and an operator. They each have their own separate business but in respect of a specific farming venture they work together. Each has an agreed share of the expenses and receives an agreed share of the income.

Tables 9.2 provides a summary of the findings from the survey on the amount of land held under different arrangements.[20] In 2006, the DEFRA survey reported 37.7 per cent of the agricultural area of England and Wales as tenanted compared to 42.3 per cent in the 2007 survey. However, as the present survey includes grass keep land (3.3 per cent) and contract farming (5.5 per cent) in the tenanted category, although technically these are not tenancy agreements, the figure from the weighted sample data is 4.2 per cent below the official agricultural samples. Our

Table 9.2   Total agricultural land area and holdings by tenure type in England and Wales in 2007 (weighted data raised to national level)

Holding type	Area (ha)	%	Holdings	%
Summary:				
Owner-occupied	6,250,319	57.7		
Tenanted*	4,577,844	42.3		
Total	10,828,170	100.0	233,745	100.0
Formal conventional:				
FAT with no share in ownership	1,891,408	17.5	33,651	14.4
FAT with share in ownership	158,999	1.5	2,716	1.2
Total FAT	2,050,407	18.9	36,367	15.6
FBT more than two years	854,152	7.9	17,432	7.5
FBT less than two years	185,790	1.7	6,326	2.7
Total FBT	1,039,942	9.6	23,758	10.2
Formal unconventional:				
Contract	595,587	5.5	2,899	1.2
Partnership	76,107	0.7	1,841	0.8
Share Farming	42,846	0.4	731	0.3
Total FU	714,540	6.6	5,471	2.3
Informal unconventional:				
Sub-tenancy	17,643	0.2	648	0.3
Grass keep	361,450	3.3	14,508	6.2
Informal/Gentlemen's agreement	271,550	2.5	15,578	6.7
Other	122,312	1.1	2,789	1.2
Total IU	772,954	7.1	33,523	14.3

*includes land held under grass keep arrangements and contract farming.

Source: Butler and Winter, *Agricultural Land Tenure in England and Wales*, 2008.

ratio of land under full agricultural tenure and full business tenure is almost identical to DEFRA's data for 2006.[21] While Full Agricultural Tenancies and Farm Business Tenancies predominate in terms of tenanted farm area (28.5 per cent of the weighted data), a significant minority of land (13.7 per cent) continues to be held under 'unconventional' arrangements as Table 9.2 shows. Furthermore, this area is actually 3.3 per cent more than was recorded in the survey of 1990, suggesting that FBTs have not solved the problem of farmers and landowners requiring flexible arrangements outside the terms of holdings legislation. Data from the RICS 2007 survey gives a percentage for sharefarming and contract farming together of 1.5 per cent of holdings, which is lower than the ADAS figure of 3.41 per cent. However, the proportion is significantly higher if looked at in terms of land area, at 5.9 per cent in Table 9.2.

The categories in Table 9.2 carry the risk of either double counting or under-recording certain categories of both tenure and occupancy. In particular, contract farming and sharefarming arrangements will take place on land that also has a tenurial status of some sort. For example, Farmer A, as owner, may be working in a sharefarming agreement with Farmer B who has no share in the land ownership but provides labour and machinery. The RICS questionnaire clearly specifies that tenure should take precedence in how land should be recorded in these circumstances. Thus the questionnaire would, if interpreted correctly, direct Farmer A to record the land as *owner-occupied* and Farmer B as *sharefarmed*. That being the case, the raw RICS data as presented in Table 9.2, as with the ADAS data, probably under-estimates the full extent of modern-day sharefarming, although we cannot rule out the possibility that Farmer A would have entered his land in the sharefarming or contract farming category, such are the challenges of using self-completed surveys to cover complex issues such as tenure and occupancy. In anticipation of this problem the RICS survey asked a supplementary set of questions on sharefarming alone to which we now turn.

The results of these questions, which have not been analysed in the reports published by the RICS, show a higher proportion of sharefarming occupancy than that recorded in the main tenure section of the questionnaire. According to the data, in England, 3.6 per cent of farmers claim to be involved in sharefarming, which is about the same as the ADAS figure. Nearly 84 per cent of farmers with sharefarming agreements own land, while over 72 per cent rent land. Table 9.3 shows that sharefarming agreements are statistically more likely to occur on farms of 200 hectares or more, which is consistent with the data already presented on the land area covered by sharefarming. Indeed the average size of a farm with a sharefarming agreement is 291 hectares suggesting that sharefarming is primarily the strategy of larger farm businesses seeking ways to expand the size of their enterprise. In analysing the

Table 9.3   Percentages of farm size categories with a sharefarming agreement

0 ha	*< 20 ha*	*20–49 ha*	*50–99 ha*	*100–199 ha*	*≥ 200 ha*
8.8	3.7	1.0	1.1	2.2	9.4

Source: Agricultural Land Tenure Survey, Centre for Rural Policy Research, University of Exeter, 2007.

Table 9.4 Percentages of tenure types and land management agreements held by farmers with sharefarming agreements

Tenure type	%
Full agricultural tenancy with shared ownership	34.9
Full agricultural tenancy without shared ownership	7.0
Farm business tenancy of less than two years	37.2
Farm business tenancy of more than two years	37.2
Contract farming	25.6
Partnership	14.0
Sub-tenancy	0.0
Grass keep	9.3
Gentleman's agreement	32.6
Other	7.0

Source: Agricultural Land Tenure Survey, Centre for Rural Policy Research, University of Exeter, 2007.

types of other agreements that sharefarmers have, Table 9.4 shows a tendency for sharefarming to be associated with a wide range of other tenure and land management agreements, particular Full Agricultural Tenancies, Farm Business Tenancies, contract farming and gentleman's agreements. If the number of agreements that a farm holds is examined statistically, sharefarmers have twice as many tenure agreements as farmers without share agreements. Looking at specific forms of land tenure and management agreements, sharefarmers have four times as many contract farming agreements, and nearly three times as many gentleman's agreements. It is highly likely, therefore, that contract farming and gentleman's agreements in some instances are close to our understanding of sharefarming. But, of course, the data is not strong enough to say how often.

Grass keep is the only form of tenure for which sharefarmers hold fewer agreements than the mean. In this case, twice as many agreements are held by non-sharefarmers as are by those with a sharefarming agreement. This is not surprising since sharefarming agreements are more numerous among arable farmers, and are statistically more likely to be associated with this type of farming as Table 9.5 shows. This is reflected in the spatial distribution of sharefarming agreements (Table 9.6), with none observed in Wales, and those in England being predominantly, but not exclusively, in arable counties.

As well as the tendency of sharefarmers to hold more varied forms of tenurial agreement they have been more dynamic in terms of land acquisition over the five years prior to the 2007 survey. For example,

Table 9.5   Percentages of main farm types with a sharefarming agreement

Dairy	Livestock	Arable	Pigs/poultry	Other
2.5	2.6	6.9	5.0	2.4

Source: Agricultural Land Tenure Survey, Centre for Rural Policy Research, University of Exeter, 2007.

Table 9.6   Counties with farms participating in sharefarming agreements

	Farms with sharefarming agreement	Do not sharefarm	Farms surveyed	% with sharefarming agreements
Oxfordshire	4	12	16	25
North Humberside	3	12	15	20
Gloucestershire	6	28	34	18
Norfolk	5	24	29	17
Buckinghamshire	2	10	12	17
Essex	3	21	24	13
Berkshire	1	9	10	10
Hampshire	2	19	21	10
West Midlands	1	11	12	8
Surrey	1	12	13	8
Hertfordshire	1	13	14	7
Kent	2	27	29	7
Shropshire	2	27	29	7
Warwickshire	1	14	15	7
East Sussex	1	15	16	6
West Sussex	1	17	18	6
Cambridgeshire	1	18	19	5
Dorset	1	22	23	4
Lincolnshire	1	27	28	4
Somerset	1	37	38	3
Cornwall & the Isles of Scilly	1	45	46	2
North Yorkshire	1	62	63	2
Devon	1	87	88	1
All other counties	0	573	573	0
All counties	43	1,142	1,185	3.63

Source: Agricultural Land Tenure Survey, Centre for Rural Policy Research, University of Exeter, 2007.

23.1 per cent of farmers in England and Wales have increased their area of land farmed over this period. However, nearly twice as many (41.9 per cent) farmers with sharefarming have increased the area of their farming operations. This trend looks set to continue with a higher proportion of sharefarmers expecting to increase the size of their farming operation; in the next five years 37.2 per cent of farmers with sharefarming agreements expect to increase the area of land that they farm compared to 16.8 per cent of all farmers. These data suggests that sharefarming agreements are a business strategy for larger farms to extend their farming operations, and frequently form part of a portfolio of land tenure agreements. Indeed, farming profitability has been a factor in increasing the area farmed in the past five years by farmers with sharefarming agreements.[22] This trend has to be seen in the context of a high level of land acquisition by non-farmers in recent decades primarily for residential or lifestyle reasons but also for investment purposes. Share and contract farming arrangements are not only favoured as a means to extend farming operations by fully commercial outfits but may also be a preferred option for land owners who have a 'lifestyle' wish to be involved in farming and in relationships with farmers that go beyond the conventional, and rather distant, landlord-tenant relation. Sharefarming may also appeal to farmers moving towards retirement who do not wish to relinquish their direct involvement in farming completely. Thus it is that 'farming to halves' may, in some circumstances, come to fulfil a new role in the complex changing social relations of agriculture.

Nonetheless, the majority of the sharefarming sample (82.9 per cent) cited financial reasons as the primary reason for having a sharefarming agreement, while 14.6 per cent cited the attraction of shared responsibility and decision-making as the prime motivating factor. In terms of what is shared, 75.7 per cent of farmers cite the sharing of farming operations as a reason for having a sharefarming agreement. All other reasons, such as sharing livestock and day-to-day management, are much less important (Table 9.7). Discussion between the two parties of sharefarming plans occurs most frequently between once a month and once every six months (Table 9.8), while a slightly larger proportion prefer meeting less often.

Table 9.7   Percentages of shared activities

Farming operations	*Day-to-day management*	*Nature conservation*	*Building maintenance*	*Livestock*	*Machinery*
75.7	16.2	10.8	5.4	16.2	13.5

Source: Agricultural Land Tenure Survey, Centre for Rural Policy Research, University of Exeter, 2007.

Table 9.8 The frequency with which plans about sharefarming are discussed

> yearly	> six monthly, but < yearly	> monthly, but < six monthly	> weekly, but < monthly	< weekly
17.5	25.0	37.5	15.0	5.0

Source: Agricultural Land Tenure Survey, Centre for Rural Policy Research, University of Exeter, 2007.

The basic share proportions between the owner of the land and the farmer are usually assumed to be in the ratio 50:50, yet we have seen from the historical evidence, and from the evidence in New Zealand that proportions could vary from this.[23] The evidence from the 2007 survey shows that the modal ratio was indeed 50:50, but that only some 29 per cent of agreements were in this modal group. Some 26 per cent of agreements had a 60:40 split between the farmer and the land owner, while the remainder ranged from 85:15 through to 15:85.

Finally, some mention is required of grass keep. Although there has been a long tradition of recognizing shared livestock ownership as an important element in 'farming to halves', grass keep is usually seen as merely a form of short-term tenure. Yet the geographical separation of sharefarming and grass keep might suggest that in the pastoral north and west, where sharefarming is infrequent, grass keep might perform some of the same functions. In one crucial respect, this is not obviously so. There is no immediate sharing of risk in traditional grass keep arrangements. Indeed the owner, or seller, of the grass as with the traditional landlord, is immune from the day-to-day risks and vagaries of agriculture. At the beginning of the grass-keep season, which usually lasts from the 1st April until the end of October, a price is agreed, often at an auction sale, and the owner of the grass receives that price whatever transpires during the grass growing season. However, there is a sharing of production costs which goes well beyond the conventional fixed costs, such as buildings, borne by the landlord. The seller of grass keep has a role in farm production in that he or she is responsible for winter and early spring grassland management so as to present the grass to the maximum advantage on sale day. The application of fertilizer and pasture management, such as rolling and chain harrowing, form part of this process, as does the maintenance of stock-proof fences. In many cases, especially with retired farmers, the seller of grass keep will play a role throughout the season in checking the stock of the farmer who has bought the grass keep. In such circumstances, especially given that the success of such an arrangement will influence its continuity and future sale prices, grass keep sales can come close to a form of sharefarming.

The purpose of this chapter has been to attempt a quantitative assessment of the extent of sharefarming in England at the turn of the twentieth century, a task that eluded us in the historical chapters. Yet, even with the benefit of modern data, government records and refined statistical techniques, we experience the familiar difficulties of under-recording, ignorance and a lack of understanding of the concept. Based on the evidence of Tables 9.1 and 9.2 we can reasonably assume that the percentage of land in England farmed by *unconventional* agreements amounts to about 10 per cent of total and that the figure is rising. As it happens, this conforms to Overton's estimate of Cornish probate inventories involved in some kind of sharefarming in the seventeenth century.[24]

10
Conclusions

In the foregoing chapters we have seen that forms of sharefarming have existed throughout the British Isles from the earliest times, and that in many cases the actual state of land occupancy and tenurial arrangements has been concealed. In all the periods we have looked at, sharefarming agreements remain elusive and are mostly concealed. As we saw in Chapter 9, even with the benefit of modern data, government records and refined statistical techniques, we experience the familiar difficulties of under-recording, ignorance and a lack of understanding of the concept. The principal problem is that the practice in England was never institutionalized, properly defined, or framed in law, unlike the situation in mainland Europe. Until the late twentieth century it remained a custom, moulded by local circumstances, known by different terms and existing in endless variation often as a verbal or gentleman's agreement. In this way it eluded the attention of manorial stewards, taxmen, lawyers, just as it was meant to. For alongside its lack of legal status, there has been the explicit desire of landowners and farmers to keep such arrangements secret and away from prying eyes. This is as true today as in the medieval period. In our experience if farmers are asked about sharefarming agreements they will more or less tell you to mind your own business; if they do oblige the information is hedged with caveats and the requirement not to divulge names or identify farms. This is hardly surprising, as confidentiality and trust have always bound these agreements. Thus, we can be sure that the full story of sharefarming in England will never be told: it will always remain largely hidden from view.

Another major difficulty has been the problem of definition. What precisely are we talking about? The taxonomy developed in Chapter 2 is largely devised from the historical experience before the twentieth century, yet, significantly, it closely resembles the six-fold taxonomy produced independently by DEFRA in 2007.[1] 'Sharefarming' and 'Contract Farming' in the DEFRA version conform to the farming to halves, profit-sharing and the modern share and contract farming in our taxonomy, while the three categories that involve the sharing of labour and machinery compare closely to the pooling and sharing of crops and resources. We have no category for 'Contract Rearing' on fixed rates per head, which we know as 'at tack', having only included ewes at halves and stock at half crease. DEFRA offers no equivalent to corn rents and agreements, but this type of support was covered by subsidies on corn production, a principal feature of the Common Agricultural Policy. The similarities between the two taxonomies, based on different evidence and using very different techniques, indicate that similar forms of sharefarming have been used by landowners and farmers consistently from the medieval period to the present day.

A significant theme running through the historical chapters was the existence of sharefarming at distinct social levels; the recent RICS survey also found this to be the case. As we saw, landlords resorted to the practice when times were hard, situations difficult or when tenants were unwilling to pay fixed rents. Amongst the peasantry, between tenant and sub-tenant, and between small owner-occupiers and labourers, it was used in a variety of ways: to overcome a temporary difficulty, ease the flow of capital and credit, expand the business and provide for old age. In its flexible way, it greased the wheels of rural society. Sharefarming in England occurred 'vertically' between landlord and tenant, and 'horizontally' amongst the peasantry, and later small owner-occupiers, conforming to the European pattern. As in Europe, fixed rents and sharefarming often co-existed on the same holding, with large tenants paying fixed rents for large farms and subletting parcels of land to halves. Similarly, a tenant might pay a fixed rent for some land, share other land to halves, hold ewes at halves or lease a herd of cows to a dairyman. This mixture of arrangements, by which landowners and farmers held their land in a variety of ways for all sorts of reasons, is an important feature of the RICS survey, underlining the growing flexibility of land tenure in the countryside and the increased visibility of unconventional agreements. An indication of such arrangements came in the British radio serial, *The Archers*, which in 2008 devoted an entire edition to a profit-sharing agreement whereby after some negotiation, a landowner gave 66 per cent of the profits to two collaborators in return for one taking responsibility for the arable on the farm, and the other responsibility for the livestock and fruit.[2]

As to the spatial distribution of sharefarming in England, we found examples in 28 counties compared to 23 in the Agricultural Land Tenure Survey of 2007. There was overlap in 16 counties, with Norfolk in fourth place, but with only Shropshire listed for the Welsh border counties. The only counties not mentioned in either survey are Westmoreland, Durham, Staffordshire, Derbyshire, Nottinghamshire, Huntingdon and Rutland. Given the covert nature of the practice these surveys remain tentative, but they indicate that forms of sharefarming existed, and still exist, in most parts of England.

The chronological incidence of sharefarming does seem to be linked with periods of depressed prices, such as the late seventeenth, late nineteenth and late twentieth centuries. So, was it, and does it remain, simply a device used by landowners and farmers in times of difficulty? The answer is no, not entirely. We saw that farming to halves was used in west Norfolk in the first half of the seventeenth century, when prices were still buoyant, as a way of establishing tenants on newly drained marshland farms, cultivating the infertile brecklands, and as an incentive to shepherds taking on newly enlarged sheep flocks. Likewise, in the Welsh Borders in the expansive sixteenth century, sharefarming developed as a way of building up capital and livestock, allowing hill farmers in this remote region to participate in the growing markets for wool and meat in England. These examples firmly linked farming to halves with improvement and risk sharing in new ventures. The most striking discovery was the re-emergence of the practice in Shropshire in the early twentieth century as family farmers, taking advantage of the newly built railway system, responded to the challenge of a growing demand for meat. Certainly, farming to halves was most evident in periods or situations of heightened risk, but this did not always correspond to depressed prices. The practice increased in the late twentieth century, when prices slumped and new laws worked against the interests of landowners. But at the same time, farmers in the RICS' 2007 survey cited the desire to expand their business as one of the principal reasons for sharefarming.

Another commonly held supposition is that sharefarming is associated with isolated areas and backward farming. The connection with Norfolk undermines this assumption. Nevertheless, it was certainly prevalent in the Highlands and Islands of Scotland, Gallic Ireland, and as we have seen, in the Welsh border counties. The link is the association with risk. Norfolk with its marshes, fens and breckland, included marginal areas which required huge investment and commitment before farms were brought into profitable cultivation. There was a long transitory phase in the seventeenth and early eighteenth centuries before tenants had sufficient capital to shoulder the entire risk of farming. As we saw on several Norfolk estates, farms which had been improved and leased to large tenants in the 1630s and 1640s, failed in the 1660s and 1670s, as corn prices collapsed. To salvage this situation, landowners resorted to farming to halves, while tenants, to help pay the rent, sometimes sublet to halves. We know this happened on leading Norfolk estates well into the eighteenth century, before they were rescued by high prices and a supportive political regime. It comes as no surprise that Norfolk landowners today, pressed by turbulent prices and a tax regime that favours joint ventures, increasingly turn to contractors to farm their estates on profit-sharing terms.

From this brief discussion we can suggest that sharefarming in England is a hidden institution which existed to supplement the working of more rigid systems of land tenure enshrined in law. These informal arrangements came to the surface in times of crisis or particular need, most notably in the fifteenth, seventeenth and twentieth centuries, as landowners and farmers required more flexibility, only to disappear from sight as the situation changed and the need receded. This flexibility demonstrates the resourcefulness and shrewd intelligence of farmers in all Ages, in contrast to the familiar stereotype of farmers resisting all change and stubbornly adhering to old routines even though circumstances might have changed radically. In the present phase of sharefarming, as the practice is supported by European and government policy, it is being defined and categorized, making the final transition from local custom to formal institution.

The growing convergence with European policy raises the question as to why sharefarming was comparatively rare in England, yet dominant in other parts of Europe. Despite our discoveries, sharefarming seems less common in England than in Europe, and indeed than in other parts of the British Isles. However, given the informal and covert nature of the practice in England, we cannot be sure.[3] What seems certain is that the English experience diverged from the European model in the early eighteenth century, as English landowners moved firmly towards a system of large farms and fixed rent tenancies; between 1718 and 1901 we found no evidence of farming to halves, in marked contrast to the experience in France and elsewhere. The reason for this difference, as we discussed in Chapter 6, appears to be largely political and institutional. For while the French, Spanish and Portuguese laboured under systems of absolute monarchy, the English landed elite secured unparalleled political power at the end of the seventeenth century.

This meant they were able to shape political, financial and legal institutions to suit their own ends. From that time they enjoyed absolute property rights, controlled levels and methods of taxation, altered the laws of mortgage to protect their estates, and passed laws to facilitate the enclosure of open fields, commons and waste.

While political power allowed the landed interest to create great personal wealth, it also removed the risks surrounding landownership. Land became a safe and lucrative investment; it also attracted a high social premium as it underpinned political patronage and power in the countryside. With a restricted electorate, landowners were able to buy and control parliamentary seats. Thus land acquired a value quite apart from the returns from agriculture. In this context, landowners did not need to maximize the returns from their estates, and could afford to concede rent reductions, invest heavily in improvements, and opt for a system of fixed rent tenancies. Above all they wanted a guaranteed income to finance their costly and increasingly cosmopolitan lifestyle. In return for paying fixed rents and assuming the risks of farming, tenants received every incentive with low rents, and the opportunity to make good profits and build up their working capital. In times of difficulty, they knew landowners would step in with abatements and investment. The benefits of this arrangement, with capital pouring into agriculture, became the envy of Europe. However, from a modern perspective, we can see that the English landlord-tenant system, effectively subsidized tenants and English agriculture. While not sharefarming in the strict sense of the word, it constituted a most generous form of risk sharing.

This system dominated English tenurial relationships, despite the Great Depression of the late nineteenth century, until the early twentieth century. In 1908, 88 per cent of the land of England was still leased to tenants on fixed rent tenancies. However, the election of a Liberal government changed the situation fundamentally. With the imposition of death duties and high levels of taxation on rental income, the loss of political power and status, and land reform promoting the interests of tenants, landownership became an increasingly heavy burden. Many landowners took advantage of the recovery in prices, which prevailed until 1921, and sold their estates to sitting tenants. Bowing to the inevitable, they actively encouraged owner-occupation amongst their tenantry, and often became owner-occupiers themselves. In the process of adjustment, we saw that forms of sharefarming played a familiar role, supplying credit, facilitating low cost expansion and sharing the risk of new enterprises. Later, the practice ameliorated the worst effects of punitive taxation and anti-landlord legislation. The principal feature of land tenure in England in the twentieth century has been the decline of the landlord-tenant system, the rise of owner-occupation and the growing popularity of short-term, flexible profit-sharing contracts. In this way, England is returning to European practice from which it diverged in the eighteenth and nineteenth centuries.

Viewed from this new perspective the English landlord-tenant system, famed for its role in underpinning a revolution in agriculture, is the exception to the rule, a passing phase rather than the cornerstone of English land tenure. This gives a new and dramatic twist to the whole story of agrarian capitalism in England. Far from being the cause of English success it was the result of a set of favourable circumstances which allowed landowners to adopt a generous system of fixed rent tenancies. Such a system, as Adam Smith pointed out, relied on optimal conditions to work effectively; when these evaporated, landowners and farmers returned to the more flexible and cost efficient methods employed by their forbears in the seventeenth century. The history of sharefarming in England should be seen in this context. As in Europe, it was a well established way of handling risk in farming, coming to the fore in the seventeenth century as landowners and farmers responded to the demands of an increasingly commercial and capitalist agriculture. In England it disappeared in the eighteenth century when landowners, secure in their wealth and political power, delegated the business of farming to tenants. However, they continued to share the risk of farming. Thus, we can see that the risk sharing model of the tenancy ladder, put forward by Offer and others, whereby risk shifted from landowner to tenant is misleading.[4] Risk sharing should be regarded as a continuous variable across contractual forms; it was not confined to particular contracts, like sharefarming, but existed in all other forms. Risk sharing was simply handled differently by different societies for various reasons to meet particular needs at certain times.

When Arthur Young dismissed *métayage* in the 1780s, the English landlord-tenant system was at its most effective. It was the ideal system for that time, sustained by high prices and a supportive government, generating massive capital investment in agriculture and expanding production sufficiently to feed a rapidly growing population. However, within a generation its shortcomings, like those of the Common Agricultural Policy in our own time, were self evident. Vested interests resist change, but eventually governments are forced to modify their institutions. This happened in England in the early twentieth century, and it partly explains the return of farming to halves after an absence of two centuries. However, it has taken another century for this story of sharefarming to emerge. Such is the extent to which Arthur Young has distorted our understanding of the development of English agriculture and the relationship between landlord and tenants, and farmers in general.

Appendix 1   A Survey of Sharefarming and its Variants in England

This is by no means a definitive survey of sharefarming in England. It is simply a summary of the examples we have detected, listed by county. Each example is categorized according to the taxonomy in Table 2.1. For the medieval period only the *champart* rents (Type A) have been included. The fixed payments in kind, which were commonplace at that time, do not appear in the list before the seventeenth century, when they were sometimes used as an alternative to the more usual fixed money rents. The examples from Scotland, Ireland and Wales have not been listed. However, the evidence suggests that forms of sharefarming were more common in outlying parts of the British Isles. Full references to published material will be found in the Bibliography.

Date	Key	Description	Source
BEDFORDSHIRE			
1294	A	At Cranfield, Elyas de Bretendon agreed to cultivate land for ½ the crop.	Raftis (1964) p. 44.
1288	A	At Shillington, Amica Atewode, the tenant of ½ a virgate, agreed that John Hammond 'sowed for ½ the crop for year to year'.	Raftis (1964) p. 76.
BERKSHIRE			
1602	A	John Laurence had 10 acres, sowed to halves in Long Wittenham field.	Berks. RO D/A1/92/133.
1610	A	Robert Loder at Harwell, 'the sayd land and fallow I put forth to halves'. Several further references to halves.	Fussell (1936).
1675	A	William Bowldry of Streatley: the corn sown to halves included 9 acres of wheat at halves and 7½ acres of barley at halves.	Berks. RO D/A1/46/112.
1675	G	E. S. Cox of Cowlease Farm, Shrivenham.	Museum of English Rural Life, SOM 6.
1880s	C	Lockinge Estate: Lord Wantage introduced allotments, co-operatives and profit-sharing schemes for his workers.	Grey (1891).
1915	E	Partnership between 2 farmers running a milk round in Newbury.	Museum of English Rural Life, BER/20/1/3.

Date	Key	Description	Source
BUCKINGHAMSHIRE			
1651	A	Letting to halves agreement between John Duncombe and Thomas Hughes at Bottle and East Claydon.	Broad (2004), p. 65.
CAMBRIDGESHIRE			
c.1300	A	At Cottenham, widow Maria Buk bound her son to take on responsibility for her holding for a half share in all the crops.	Ravensdale (1984).
1589	A	At Willingham, the will of Richard Biddall directs his brother to manage 2½ acres for his son for half the profits.	Spufford (1974), p. 140.
1872–1874	C	Lord George Manners of Cheveley Park experimented with partnership farming 1872–74, dividing profits of Ditton Lodge Farm among his labourers.	Lewis (2002), p. 92. *Cambridge Chronicle*, 10 Jan 1874, p. 8; 19 Sept 1874, p. 7; 12 Dec. 1874, p. 4.
CHESHIRE			
1656	A, G	Sowing to halves and cow leasing: from the Memorandum Book of Thomas Jackson of Hield, Aston by Budworth, 1622–1707.	Foster (2002), p. 92.
CORNWALL			
1606–1735	A, B, E	References to half crease, moieties and halfendeale in probate inventories.	Cornwall RO.
1659	E	St. Just in Penwith, Martin Ling had a moiety of 2 fields worth £3 10s.	J. Thirsk, pers. comm.
1659	E	St. Martins, Henry Hoskins, had a moiety of a living £10.	J. Thirsk, pers. comm.
1662	E	Feock, R. May had ½ shares in 2 kine, 2 calves, 1 heifer, 2 mares, 1 nag, 7 pigs, poultry and geese, 6 acs of wheat, 7 acs of barley & oats.	J. Thirsk, pers. comm.
1815	G	Description of cow leasing.	Worgan (1815), p.141.
1728	G	Cow leasing at Liskeard, profit from a cow as high as £3 10s, or £4.	PRO E, 1 Geo II, Easter 3.
1960s and 1970s	B	Half crease at Liskeard whereby farmers allowed their land to be stocked by other farmers; common practice on Cornish family farms.	J. Lewis, pers. comm. C. Langley, pers. comm.
CUMBERLAND			
1590s	E	J. Purchase, clerk of Ireton has 'some 3rd part of sheep'.	J. Thirsk, pers. comm.

Date	Key	Description	Source
1610s	E	E. Nicholson has 20 sheep at 3 parts & 24 sheep at 3 parts with 2 other men. People share bees. One has ½ a hive in J. Grave's garth.	J. Thirsk, pers. comm.
1630s	E	Fells, Ann Vepond has £1 worth of sheep in half part & 17 ewes, 1 tup, 17 hogs.	J. Thirsk, pers. comm.
	A	Suggesting use of farming to halves.	Dilley (1970), p. 192.

DEVON

1952	A	Lambs at half crease.	Stanes (1990), p. 7.
1679	G	Cow leasing at Colyton, at *usual rate* of £3 10s per cow.	Harrison (1984), p. 378: PRO, E 134, 31 Chas II, Easter 3.
1697	G	Cow leasing at Lapford, lesser area than Colyton, rate 24s, 25s.	Harrison (1984), PRO, E 134, 13 Wm Trin. 7.

DORSET

1712	G	Cow leasing at Mappowder. Daniel Holland rented out his dairy of 24 cows at £3 a cow.	Harrison (1984), PRO, E 134, 13 Anne, Easter 10.
1756	G	Dairy agreement for Corfe Castle, Dorset, rental £3 a cow.	Harrison (1984), p. 378, fn. 137, Dorset RO, P11/IN5.
1758	G	Cow leasing by James Warne at Bovington.	James and Bettey (1993).
1781	G	'Rec'd of Mr James a qtrs rent for 5 cows' at Poxwell.	Museum of English Rural Life, DOR, 8/1/1.
1794	G	Cow leasing described. Rents varied between 50s and £3 a head in poorest parts. Up to £7 in good areas. Broad Windsor, Beaminster £8.	Claridge (1794), pp. 14–15.
1815	G	Cow leasing described further, building on Claridge above. Rents rose steeply during French wars.	Stevenson (1815), pp. 381–90.
	G	Examples of cow leasing from early 18th century to late 20th century.	Horn (1978), pp. 100–107.

HAMPSHIRE

1813	G	Cow leasing described in County Report. Cows rented to dairymen at £7 to £9 a cow.	Vancouver (1813), p. 364.

HEREFORDSHIRE

1347	A	Demesne lands of Mortimer family at Wigmore, for a ⅓ sheaf.	Hilton (1990), p. 513.

Date	Key	Description	Source
KENT			
1604	A	John Austen of Northbourne, yeoman, agreed with Simon Lott, husbandman, that he should sow to halves land Austen leased from Sir Thomas Peyton.	Chalklin (1965), p. 63.
c.1650	A	5th Earl of Dorset accorded 4 farmers the right to 'plough anywhere in the Park … ' in return for ⅓ of the crop. 'The third year the farmers to sow the ground with grass seed, they are to be at the charge of seed, tillage and harvest.'	Venn (1923), pp. 43–8.
1620–99	A	References to halves in probate inventories	Centre for Kentish Studies
LANCASHIRE			
c.1650	A	Cows to halves.	BL Add MS 33509. fol. 74; Speed (1659).
1622	A	Walmesley family put to halves lands at Dunkenhalgh.	Lancs. RO, DDPt/19 (agisters' accounts 1612–20).
LEICESTERSHIRE			
1315	A	Duchy of Lancaster, manor of Desford, *champart* lease to group of manorial tenants.	Hilton (1990), p. 513.
LINCOLNSHIRE			
1656–1683	A, G	Drayner Massingberd of South Ormsby cum Ketsby. Account Book lists. Several agreements with small tenants and labourers which include elements of share cropping. He also leased them sheepgates and cowgates, for the grazing of small numbers of animals.	Holderness (1972).
1660s	A	Robert Toppin for agreeing to sow arable to halves.	Lincs. RO, MM/VI/1/5.
1668	A	W. Wikam agreed to sow to halves part of the Great Close – 'same manner as Robert Toppin doth the other part'.	Lincs. RO, MM/VI/1/5.
1683	A	Agreement between Drayner Massingberd and William Brookes of Calceby, labourer, where DM provided all stock, seed, WB cultivated corn, looked after sheep, repaired fences, etc. in return for 8th part – later altered to 6th part.	Massingberd (1893).

Date	Key	Description	Source
MIDDLESEX			
c.1300	A	Westminster Abbey demesne at Yeoveney, *champarty* lease.	Hilton (1990) p. 513.
NORFOLK			
1270	A	Methwold, lease of half an acre, *ad campi partem*.	Homans (1940), p. 202.
1420	A, H	Pastons at Edingthorpe sowed to halves and used barley rents on their east Norfolk properties in 1470s.	Norfolk RO (NRO), PHI 532 578 X 2.
1587– 33	A	References to 'ptable' crops in probate inventories.	NRO.
Hunstanton Estate			
1605 1622	A	'The profit of the tithes and halfes of the Corne growing at Barrett Ringstead. Foldcourse let to several tenants, payments for 'my halfes'.	NRO, LEST/R8; NRO, LEST/Q38; NRO, LEST /P6. NRO, LEST/BK3; NRO, LEST/P6.
1613	A	Sowing to halves agreements with Thomas Ketwood and William Cobbes, for closes near Hunstanton Hall.	
1637– 1644	A	Sowing to halves agreements with tenants on newly drained marshland at Hunstanton and Holme.	NRO, LEST/KA6; KA9; KA24; NRO, LEST/P7.
1669– 1683	E	Trustees of Sir Nicholas Le Strange of Hunstanton, Account book, includes refs to cow grasses, halving flocks and foldcourses.	NRO, LEST / KA11.
Raynham Estate			
1470– 1485	A	Raynham estate: sharecropping at Stibbard and Lt. Ryburgh.	Moreton (1992), pp. 146–7.
1622– 1637	A	Payment of 'partes' or shares of wool and lambs to shepherds for the flocks at Raynham, East and West Rudham, South Creake, Barmer, Barwick and Shereford.	Townshend MSS (TM) Sheep Reeve's accounts, RAS/F2/11.
1634	A	William Grove 'sowed to halfes' 3 acres in West Rudham.	TM, RAS/F2/12.
1637	A	Cooper 'sowed to halfes' 2 acres at Kipton.	BL Add. MS 41308.
1650s	A	At Kipton, 'the brecks in halfes' with Mr Stringer.	TM, RAD/A5.
1665	E	Division of the brecks into 3 parts at Great Grounds, West Rudham; and foldcourses leased to partners.	TM, RAD/A4.
1667	A	At Raynham, Sam Jervise sowed to halves the Lords brecks and his own infield lands with the Lord.	TM, RAD/A4.

Date	Key	Description	Source
1679	A, G	Raynham Hall dairy leased, Timothy Felton recommends 'putting to halves' meadows in the park.	TM, RAS/A1/5.
1682–1690	A, G	At South Raynham, W. Downham's Bill of Sale and agreement for 1 year 'in money and half the corn crop'. J. Raby leased 6 cows and cultivated the arable.	TM, RAS/A1/6.
1677–1686	A	R. Teasdale letting to halves agreement for Stiffkey Hall Farm; 1677–1684. New agreement proposed 1686 not implemented.	BL Add. MS 41655/160.
1684–1698	G	Stiffkey Hall dairy of 20 cows set up and leased to Nick Cooke, who cultivated the arable. 2 Foldcourses in hand.	TM, RAS/A1/6 ; RAS/A2/1–8.
1689–1698	G	Langham Farm, J. Raby leased 13 cows and cultivated the arable, with a man for £28 a year. Foldcourse in hand.	TM, RAS/A1/7.
1690–1697	A	East and West Rudham, 3 Foldcourses in hand, arable let to halves.	TM, RAS/A2/1–8.
1691–1698	A, G	Toftrees estate, foldcourse in hand, arable let to halves, dairy leased.	TM, RAS/A2/1–8.
1692	A	South Creake, 2 foldcourses taken in hand, arable let to halves.	TM, RAS/A1/6; RAS/A2/1–8.
1706	G	Shipdham Estate: tenant subletting 4 dairies.	BL Add. MS 41656/ 182–183.

Felbrigg Estate

Date	Key	Description	Source
1662	A	Dilham Hall Farm broken up and let to halves for 1 year.	NRO, WKC, 5/422 464 X 4.
1677	A	Letting to halves agreement for improving Rush Close, Gresham for 13 year term – ran its course.	NRO, WKC 5/152 400 X; Griffiths (2000).
1677	A	Agreement with Thomas Sexton for Reepham Farm for 5 years.	NRO, WKC 5/152 400 X.
1678	A	Agreement for the Park Dairy and Grounds, Felbrigg, ended 1681.	NRO, WKC 5/152 400 X.
1679	A	Agreement with J. Fincham for Selfe's Farm, Felbrigg, ended 1681.	NRO, WKC 5/152 400 X.
1681	A	At Dilham, J. Applebye agreed to sow Mack's farm to halves for 1 year.	NRO, WKC, 5/152 400 X.
1682	A	Letting to halves agreement for Parke's Farm, Alby, ended 1684.	NRO, WKC 5/152 400 X.
1689	G, H	Beckham Hall, John Lound leased cows and paid rents in corn.	NRO, WKC 5/158 400 X 6.

Date	Key	Description	Source
Blickling Estate			
1666	G	Plan for Langley Abbey involving leasing of a dairy.	NRO, NRS 10379 25 A 6.
1677	G	Agreement with John Cullyer of Intwood Hall, for leasing dairy and contracting the arable at fixed rates.	NRO, NRS 11321 26B5.
1679	J	Corn agreement at Horsham St. Faiths.	NRO, NRS, 16023 31 F 10.
1718	A	Langley Estate, tenant Richard Berney halved lands with sub-tenant.	NRO, NRS, 16338 32C2.
Elsewhere in Norfolk			
1666	A	Simon Britiffe let Cley Hall 'to halfes with my tenant Flaxman'.	NRO, GTN, 425.
1689	A	At Houghton, Sir Robert Walpole's accounts contain several references – 'to Will Rowardes for halfeing the plowground'.	Cholmondeley (Houghton) MSS; List of Account Books 7–19.
1698	A	At Westwick, John Berney Gt. 1 year agreement with John Ollyet, yeoman, to farm to halves the holding of a widow.	NRO, PET, 159 97 X 2.
1960– 2009	D	Profit-sharing, modern agreements based on East Anglian tradition of taking a crop, undertaken by Jim and David Papworth, Felmingham, with 30 or so partners.	J. Papworth, pers. comm.
NORTHANTS			
1235	A	Abbey of Cirencester, 2 virgates *ad medietatem.*	Hilton (1990), p. 515.
NORTHUMBERLAND			
1590s	A	Lowlands, Robert Finch has 3 sows 'at half part'.	J. Thirsk, pers. comm.
1880s	C	Albert Grey used a profit-sharing scheme on his Northumberian estate at Howick, Chevington Moor and East Learmouth.	Grey (1891), pp. 781–93.
OXFORDSHIRE			
1302	A	Launton, *champarty* lease.	Harvey (1977), p. 319.
SOMERSET			
n.d	A	Medieval Customs of the Manors of Bradford on Tone and Taunton.	Miller and Hatcher (1978), p. 143.
n.d	A	Tenant at Middlezoy, (Longleat Estate) covenanted to half the profits of 18 acres.	Postan (1973), p. 136.

Date	Key	Description	Source
1715	G	Profit of milk estimated at £2 10s a cow.	Harrison (1984), p. 378; PRO E 134, 5 Geo I, Mich 2.
1745	G	Samuel Hipsley at Bleadon, milk from cows valued at £4 a year.	Harrison (1984), p. 378; Somerset RO, D/D/C 1747.

SHROPSHIRE

Date	Key	Description	Source
1269	A	Brockton woman surrendered her land to Wombridge Priory, in return for ⅓ of the grain, and ¼ of rye, if they did not till her garden croft.	Stamper (1989), p. 37.
1295	A	Adam, from Rossall gave a *champart* lease of all his land, except 13acs otherwise leased to 2 brothers, also from Rossall.	Stamper (1989), p. 37.
1308	A	Walter of Wenlock, *champarty* agreement.	Harvey (1977), p. 318.
1270–1349	A	Court Rolls of Halesowen give 21 examples of sharecropping agreements. Table 2: Economic Activities of Halesowen Villagers.	Razi (1987), p. 367.
1375	A	Richard Moulowe of Halesowen sued Henry Townhall for ½ the crops.	Razi (1987), pp. 389–90.
1399–1468	A	Lordship of Blakemere, ⅓ sheaf oats paid to the lord, from a field previously worth £5 6s 8d; context Welsh raids in Whitchurch 1410.	Kettle (1989), p. 103.
1405	A	Sowing for 4th sheaf used in parts of the Lordship of Oswestry as a way of attracting back tenants and raising revenue after Welsh raids.	Kettle (1989), p. 103.
1367–1383	A	Lordship of Caus, land leased for ⅓ sheaf; demesne arable leased with oxen for ⅓ sheaf, preliminary to leases for years and money.	Kettle (1989), p. 103.
1552	B	John Mytton of Kinnerton has his kine and heifers set to parts/ to hire.	Kettle (1989), p. 151.
1820s	J	Lilleshall Estate, Leveson-Gower, agent James Loch agreed that half the tenants rent could vary with the price of corn. Reversed 1830s.	Kettle (1989), p. 214.
1900–1930	A	Coppice Farm, Ratlinghope, W. Cooke placed ewes at halves with neighbours.	D. Cooke, Ratlinghope Farm Diary.
1908–1940	A	Curdale Farm, Cleobury Mortimer, Evans family farmed to halves, sheep and cattle with neighbours.	M. Evans, Curdale Farm Diary.

Date	Key	Description	Source
1914–17	A	Lower Elcott Farm, Neen Savage, Benjamin Griffiths had 'sheep at halves' with Walter Amies of Kyre.	J. Griffiths, Elcott Farm Diary.
1920s–1930s	A	Edward Foster at Newton, Bridgnorth on the Apley Estate had 'sheep to halves' with another tenant nearby.	J. Bevan, pers, comm.
1940s	A	Overwood Farm, Cleobury Mortimer, Charles Griffiths had sheep at halves with friends nearby for 2 to 3 years.	J. Griffiths, pers. comm.
1950s	A	Wall Town Farm, Kinlet, George Haywood had ewes at halves with several neighbours, plus 500 ewes with a friend at Broadway, Glos.	J. Haywood, pers. comm.

SUFFOLK

Date	Key	Description	Source
1692	A	Daniel Cook, gent of Sudbury, deposed that in 1688–90 Sam Warner held 14 acres arable, sharing the lands with Stephen Carter 'according to the usuall way or manner of halving' – Warner to find the land, barn and pay taxes & Carter to plough sow and cut corn. Sam Carter deposed that Carter had told him he received half the crop.	PRO E. 134 4 William and Mary, Mich. 43 Luke Leake v. Samuel Warner, Sudbury 20th Oct 1692.

WARWICKSHIRE

Date	Key	Description	Source
1249	A	Lessee paying ⅓ produce for demesne of local priory near Coventry.	Hilton (1990), p. 513.
1411	A	Coventry Cathedral Priory, single demesne lease for a ⅓ crop.	Hilton (1990), p. 513.
1400	A	Longdon, *champarty* lease.	Harvey (1977), p. 319.

WILTSHIRE

Date	Key	Description	Source
	G	General references to cow leasing, referred to as sharemilking.	Wilson (1995), pp. 51–5.

WORCESTERSHIRE

Date	Key	Description	Source
1230	A	Worcester Cathedral Priory, Pensax, leased demesne for ⅓ sheaf.	Hilton (1990), p. 512.
c.1300	A	Worcester Cathedral Priory, Wolverley, group of tenants paid ½ crop for manured demesne, ⅓ crop for unmanured lands.	Hilton, (1990), p. 513.
1370	A	Earl of Warwick, demesne of Elmley Castle, tenants of manor paid second sheaf.	Hilton (1990), p. 513.

Date	Key	Description	Source
1343	A	Pinvin, *champarty* lease.	Harvey (1977), p. 319.
1372	B	Two prosecutions at Wolverley, for disputes concerning shared profits for sheep and cattle.	Hilton (1973), p. 43.
1372	A	Widow of Heye, tried to get a man to cultivate 2 acres for half crop.	Hilton (1973), p. 50.

YORKSHIRE

Date	Key	Description	Source
1615	H	Cowick, Edmund Ossett had ½ share of 9r of wheat & rye.	J. Thirsk, pers. comm.
1625	E, G	Several references to cow leasing, farms being halved at Elmswell, East Yorkshire.	Woodward (1984), p. 179.
1671	E	Wawne, East Yorkshire. Half shares in 4 farms were sold in 1671 by Thomas Thynne and Devereux Leicester.	Carrick (2004), p. 67.
1890s	A	Castle Howard, Lady Carlisle operated a *métayage* system, whereby occupiers paid ⅓ crop which covered her rents, rates and return on working capital.	Grey (1891), pp. 780–1.

WELSH BORDER COUNTIES

Date	Key	Description	Source
1347	A	Demense lands of Mortimer family (at Radnor) for a ⅓ sheaf.	Hilton (1990), p. 513.
1500s	A	Several references to 'half the increase' in Radnorshire. Share cropping was a key element of Cattle Economy; written agreements with penalty bonds, pursued in courts. Litigated cases.	Suggett (2005), pp. 181–210.
1500s	A	Examples of half the increase extend to other marcher counties: Montgomeryshire and Breconshire. Hiring cattle and sheep prevalent in Cardiganshire – see Cornwall for half crease.	R. Suggett, pers. comm.

GENERAL REFERENCES

Date	Key	Description	Source
1500s	A	'Letting and sowing to thirds and halves was practiced in the Cotswolds, Chiltern, Oxford Heights, Fen and Cheshire Cheese Countries, The Vale of Evesham, the Wealden Vales, the Vales of Hereford, the Midland and Lancashire plains'.	Kerridge (1969), p. 52.
	G	References to cow leasing in the west country, 1650–1850.	Broad (2004), p. 103.

Appendix 2   Letting to Halves at East and West Rudham in 1693

Evidence for letting to halves appears in three kinds of documents: the bailiwick accounts, which provide a description of the holdings let to halves; the separate 'letting to halfes' accounts; and finally, the agreement setting out the terms of the contract, attached, with the supporting vouchers, to the back of the account. The first description of letting to halves is contained in the bailiwick account of East and West Rudham for 1693. It is an extract of the lands, associated with Grannohill and Manyards foldcourses, let to halves to the two tenants, John Butler and William Roads. The detail shows the close integration of the management of the foldcourses with the brecks, heaths, outfield and infield lands, and earlier attempts to structure balanced holdings. Not all the lands were let to halves; some closes were let on fixed rents and some retained in hand. This can also be seen at Toftrees, where difficult holdings were identified for letting to halves, while others were leased on fixed rents. The entry for South Creake is a simpler affair, with the principal holding let to halves and six parcels hived off and leased to smallholders.

The letting to halves agreements do not resemble Windham's, or rather Britiffe's, complex agreements at Felbrigg. They are more akin to the corn agreements Brewster instigated at Saxthorpe and Horsham St Faith's on the Hobarts' estate, in that they were confined to corn growing. However, whereas Brewster guaranteed a minimum price to tenants, in effect offering a subsidy, the trustees genuinely shared the risk of cultivating specified acreages of corn, as can be seen in the agreements with John Butler for East Rudham. The agreement, on a postcard-sized piece of paper, was made in August, based on an estimate of the crop yields, in which the tenant agreed to deliver the half, or moiety, in the barn. John Butler agreed to sow four acres of wheat at two combs to the acre, and deliver four combs 'the moyeity' or half of the crop. These amounts were then quoted at the head of the account. The few surviving agreements for the Rudhams, Tofts Farm and South Creake, show the acreages sown to different crops and yields anticipated: wheat yields at South Creake were double those expected at Toftrees and the Rudhams, where in fact very little wheat was grown by Butler and none by Roads. The acreage appears small, but allowance must be made for lands fallowed and laid down to grass.

The 'let to halfes' accounts, settled on 20 June of the following year, quote the contract, list the deliveries of corn – and what remains due – the price made and the money received. Added to the total were the profits from rents, and sale of stock or goods. Allowances included local rates and the costs of cultivation, which were substantially lower and less complex than the returns for farms in hand at Stiffkey and Langham. The allowances for South Creake Farm include payments made to the tenant for 'fencing ye closes & meadows & clearing pte of the brecks', indicating that improvement was the primary motivation for the practice. J. Butler's account for 1693, shows Colonel Walpole acting as adjudicator. The account can be compared against the agreement.

EXTRACT 1: From E. and W. Rudham 1693, showing some land let to halves to J. Butler and W. Roads

---

Lett to halfes late in Rowland's use, Nick Close & 54acs of land at 4/6 p.ac £13.7.3 & ye rest of Bennetts lands at 4/6 p.ac at £7.13.0 but of this Barrett hath by agreement 3ac at 13/6 p.ann and formerly in Rowland's use, 10ac pte of the 54acs at 5s p.ac all in Rudham North field and lett to halfes to J. Butler, but the 10ac is lett to halfes to Wm Roads so to be accounted for by themselves with ye farmes in hand being £17.16.9 p.ann so not charged here   000 00 00

Late Wid. Russells & Fishers use 78acs at4/6 p.ac, but of this late in Rowlands use 48acs 2r & now lett to halfes to J. Butler & to be ac for wth ye farmes in hand being 29acs 2r at 4/ p.ac £6.12.9 p.ann soe not charged here   000 00 00

EXTRACT 2: J. Butler's letting to halves agreement for 1693 made in August 1692

Rudham 26 August 1692
John Butler agreed to deliver particular quantities of all sorts of corn and grayne hereunder mentioned to and for the use of the Executors of the late Lord Townshend valued as ye moyeity of the crop now growing upon the lands of East Rudham

			xx	c	b
Wheat	4 acres	ye moyeity @ 2C p. acre		4	0
Rye	40acres	ye moyeity valued at		80	0
Barley	78 acres	2r ye moyeity valued at 4C 2b p. acre	8	16	2
Oats	36 acres	ye moyeity valued at 4 C p. acre		72	0
Pease	12 acres	3r ye moyeity valued at		10	0
Fetches	12 acres	allowed for horsemeat			

Witness   Peter Stringer   J. Butler

EXTRACT 3: J. Butler's letting to halves account for 1693

The Accountant chargeth himself wth the moyeity of the Corne to be delivered by John Butler as by his agreement

	xx	c	b	£	s	d
Of wheat	0	04	0			
Of Rye	4	00	0			
Rem due last year	0	19	0	4. 19. 0		
Of Barley	8	16	2			
Rem due last year	1	04	0	10. 00. 2		
Of Oats	3	12	0			
Of Pease	0	10	0			
remain due last year	1	10	2	2. 00. 2		

Wheat
This accnt chargeth himself wth
  money rec'd for   0   4   0   of wheat @ 16/-   3   4   0

Rye

Delivered by Butler to shepherd at Manyards	0	9	2	of Rye covenant Corne				
more unto malt chamb which I must ac next yr	1	8	0					
Exd	1	17	0					
soe rem due from J. Butler	3	1	2					

Barley

Money rec'd of Butler for Exd	8	16	2	of Barley @ 9/4, 9/8, 10/- p.c	80	2	0
soe rem due from Butler	1	4	0	of Barley			

Oats

Money rec'd for	1	1	0	of Oats @ 6/- p.c.	6	0	0
Delivered to Raynham Hall for Ldships use	2	11	0	of Oats			
	3	12	0	rem nothing			

Pease

delivered none so remain due from Butler	2	0	2	of pease			
				The whole rec of Butler Corn	89	6	0

The Acctnt chargeth himself wth the moyeity of W. Roads corn growing in ye yere 92 & pte in the field & tyth in My Ld's Execs charge

	xx	c	b		£	s	d
Rye	3	16	1				
Barley	5	2	1				
Oats	3	10	1				
Rye delivered to shepherd at Grannohill	0	6	0				
malt chamber & granary	3	10	0				
	3	16	0				
Barley money rec'd of John Tid	4	14	2	of Barley @ 9/-, 10/-	43	5	0
delivered to shepherd	0	4	0				
to Raynham for fowls & pigeons	0	3	3				
	5	2	1				
Oats delivered to Raynham for Ldships use	3	10	2				

The Acctnt chargeth himself wth ye Corn of Mr Danells tyth & of land late in his use

	xx	C	B		£	s	d
Wheat	0	2	0				
Rye	0	17	1				
Barley	1	15	1				
Oats	1	12	2				
Pease	0	6	0	3 pecks			
Wheat money rec for	0	2	0	of wheat	1	16	0
Rye "	0	17	1	of rye @ 11/- p.c	8	15	0
Barley "	1	2	3	of barley	10	5	6
pigeons & fowls	0	12	2				
	1	15	1				
Oats money rec for	1	1	0	of oats 6/- p.c	6	0	0
delivered for seed for his Ldshps land at Tofts and Raynham	0	11	2				
	1	12	2				
Pease money rec for	0	6	0	3 pecks white pease £1.12.0	3	2	6
				The whole sum rec is	29	19	0

and wth the money rec of T. Denis for 1 yrs rent of a meadow	2	0	0
D. Riches for 1 yrs rent of the Malthouse	7	0	0
M. Dodman (shepherd) for 1 yrs rent of 2 acs late in Danells use	0	9	0
for straw and chafe for Danells tyth	1	15	5
for the fayre stall and things standing	1	0	0

	£	s	d
Profits	12	4	5
Butlers crop	89	6	0
Roads crop	43	5	0
Danell tyth	29	19	0
whole sum rec	174	14	5

There is due from J. Butler of corne which he did not deliver in this acc and what was rem last yr according to his agreement of Rye 3.1.2, of Barley 1.4.0 of pease 2.0.2 all which he hath acc and answered the value & his bills & disbursements as by an acct. made & stated in the presence of Col Walpole 9 Oct 1693 where by there doe appear to be due from said Butler for the corne above mentioned the sume of £87.9.6. the copy of the acc is hereinto annexed

[Similar entry for W. Roads ... £42.17.9]

Acctnt ... for money payd to T. Denis for his harvest wages ending in Roads moyeity	2	13	4
H. Mason Mr Danells tyth	3	16	6
T. Denis for threshing Mr. Danell tyth £1.17.10; moyeity of Roads crop £6.17.2	8	15	0

1 yrs rent for Wm Ramseys – 2 receipts annexed	50	0	0
Hedging, ditching, catching moles	2	7	10
Poor rate 6.3.9, constable rate 2.3.3, church rate 1.16.10 for E/W Rudham	77	16	1
Exd June 20 1693 Edwd Le Strange Debenter	96	18	4

Notes

Chapter 1

1 A. Young, *Travels through France During the Years 1787, 1788, and 1789* ed. C. Maxwell, (Cambridge: Cambridge University Press, 1929), p. 16.
2 *Ibid.* p. 287.
3 *Ibid.* p. xxvii; A.R.J. Turgot, *Reflections on the Formation and Distribution of Wealth*, (Paris, 1766), pp. 26–31.
4 M. Bloch, *Rural French Society*, (Routledge and Kegan Paul, 1966), pp. 146–9.
5 P. Goubert, *The Ancien Régime: French Society 1600–1750*, (Weidenfeld and Nicolson, 1969), pp. 108–13; J. Blum, *The End of the Old Order in Rural Europe*, (Princeton: Princeton University Press, 1978), pp. 102–3.
6 R.H. Hilton, 'Why was there so little champart rent in Medieval England?', *Journal of Peasant Studies*, 17 (1990) 509–19.
7 J. Thirsk (ed.) *The Agrarian History of England and Wales*, 8 vols, (Cambridge: Cambridge University Press, 1967–2000); M. Overton, *Agricultural Revolution in England: the Transformation of the Agrarian Economy, 1500–1800*, (Cambridge: Cambridge University Press, 1996).
8 D.R. Stead, 'Risk and risk management in English Agriculture, c.1750–1850', *Economic History Review*, 2nd series, 57 (2004) 334–61; A. Smith, *An Inquiry into the Nature and Causes of the Wealth of Nations*, (1776), Book 3, Part 2.
9 E.M. Griffiths (ed.) *William Windham's Green Book 1673–1688*, Norfolk Record Society, Vol. LXVI (2002); *idem*, 'Responses to Adversity: the changing strategies of two Norfolk landowning families, c. 1665–1700' in R.W. Hoyle (ed.) *People, Landscape and Alternative Agriculture: Essays for Joan Thirsk*, Agricultural History Review Supplement 3 (2004), pp. 74–92.
10 Goubert, *Ancien Régime*, gives an example of *métayage* at Parthenay, Deux Sevres in 1649, pp. 142–4.
11 Mark Overton now estimates that some 5 per cent of farmers' probate inventories in Norfolk, from the 1580s to the 1730s, include some reference to sharefarming, but did not pursue the implications of this while researching for his PhD in the 1970s. John Broad, working on documents long familiar to him, 'discovered' a letting to halves agreement on the Verneys' estate in Buckinghamshire dating from 1651: Verney MSS BLM636/11.
12 G.E. Fussell (ed.) *Robert Loder's Farm Accounts 1610–1620*, Camden Society, 3rd series, Vol. 53 (1936).
13 Overton, *Agricultural Revolution*, pp. 203–6.
14 A. Offer, 'Farm tenure and land values in England, c.1750–1950', *Economic History Review*, 2nd series, 44 (1991) 1–20; Stead, 'Risk and risk management', pp. 334–61.
15 M.E. Turner, J.V. Beckett, and B. Afton, *Agricultural Rent in England, 1690–1914*, (Cambridge: Cambridge University Press, 1997), pp. 6–36; Overton, *Agricultural Revolution*, pp. 30–5.

16 *Ibid.* pp. 203–6.
17 S. Ogilvie, '"Whatever is, is right"? Economic institutions in pre-industrial Europe', *Economic History Review*, 2nd series, 60 (2007) 649–84.
18 P. Hoffman, *Growth in a Traditional Society: the French Countryside 1450–1815*, (Princeton: Princeton University Press, 1996); A. Antoine, 'La Légende noire du métayage dans l'ouest de la France (XVIII–XIX siècle)', in G. Béaur, M. Arnoux, and A. Varet-Vitu (eds) 'Exploiter la Terre: Les Contrats agraires de l'Antique a nos jours, Actes du colloque international, Caen, 1997', *Bibliothèque d'Histoire Rurale*, 7 (2003) 457–70; J.M. Moriceau, 'Fermayage and métayage (XII–XIX siècle)', *Histoire et Sociétés Rurales*, 1 (1994) 155–90; A. Antoine, 'Métayage, productivité et économie monétaire. Les enseignements de l'analyse des comptabilités agricoles', *Agricultural History Review*, forthcoming.
19 Smith, *Wealth of Nations*, Book 3, Part 2; Turgot, *Reflections on the Formation and Distribution of Wealth*, pp. 26–31.
20 J. Cohen and F.L. Galassi, 'Sharecropping and productivity: feudal residues in Italian agriculture, 1911', *Economic History Review*, 2nd series, 43 (1990) 645–56; F.L. Galassi, 'Moral Hazard and Asset Specificity in the Renaissance: The Economics of Sharecropping in 1427 Florence', in K.D. Kauffman (ed.) *Advances in Agricultural Economic History*, Vol. 1, *New Frontiers in Agricultural History*, (Stamford, CT: JAI Press, 2000), pp. 177–206; Bloch, *French Rural Society*, pp. 142–4 for a comparison with Italian forms of sharefarming.
21 F.M. Snowden, *The Fascist Revolution in Tuscany 1919–1922*, (Cambridge: Cambridge University Press, 1989), pp. 7–69 for the sad tale of *mezzadria* from its golden age in the mid-19th century to its decline during the depression of the late 19th century and the crisis of the early 1920s. In 1923 the Fascists re-imposed the paternalistic model successfully challenged by the *mezzadri* in the uprising of 1920.
22 F. Ellis, *Peasant Economics: Farm Household and Agrarian Development*, (Cambridge: Cambridge University Press, 1993), pp. 146–66.
23 R.W.M. Johnson, *Reforming EU Farm Policy: Lessons from New Zealand*, (Institute of Economic Affairs, 2000); G. Blunder, W. Moran, and A. Bradly, 'Archaic relations of production in modern agricultural systems: the example of sharemilking in New Zealand', *Environment and Planning A*, 29 (1997) 1759–76; R. Stratton, A. Sydenham, and A. Bird, *Sharefarming: the Practice – With Model Form of Agreement*, 3rd edn (Country Landowners Association, 1992). In the 1980s professional advice on sharefarming contracts was more or less confined to the west country firm led by Richard Stratton, an early enthusiast with long standing contacts in New Zealand. Since then Stratton and Holborow, have been joined by the leading land agency firms such as Strutt and Parker, and JPD Savills who offer a full advisory service with specialists in Contract, Share Farming and Joint Ventures.
24 J. Thirsk, *Alternative Agriculture: A History since the Black Death to the Present Day*, (Oxford: Clarendon Press, 1997), pp. 1–4.

Chapter 2

1 J.S. Cohen, 'Institutions and Economic Analysis', in Thomas G. Rawksi *et al.* (eds) *Economics and the Historian*, (Berkeley: University of California Press,

1989), pp. 60–84; F. Ellis, *Peasant Economics: Farm Households and Agrarian Development*, (Cambridge: Cambridge University Press, 1993) pp. 146–64 for useful introductions. See also P. Schofield and B. Câmara, 'Introduction, Sharecropping in History', *Continuity and Change*, 21 (2006) 209–11.

2 A. Smith, *An Inquiry into the Nature and Causes of the Wealth of Nations*, (1776), Book 3, Part 2, p. 474.

3 A. Marshall, *Principles of Political Economy*, 8th edn, (Macmillan, 1961), pp. 535–7, makes a 'fundamental distinction between the "English" system of rental and that of holding land on shares as it is called in the New World, or the "Métayage" system as it is called in the old'.

4 S. Ogilvie, '"Whatever is, is right"? Economic institutions in pre-industrial Europe', *Economic History Review*, 2nd series, 60 (2007) 649–84.

5 D.M.G. Newbery, 'Risk sharing, Sharecropping and Uncertain Labour Markets', *Review of Economic Studies*, 2nd series, 44 (1977) 585–94.

6 J. Carmona and J. Simpson, 'The "Rabassa Morta" in Catalan Viticulture: The Rise and Decline of a Long-Term Sharecropping Contract, 1670s–1920s', *Journal of Economic History*, 59 (1999) 290–315; F.L. Galassi, 'Moral Hazard and Asset Specificity in the Renaissance: The Economics of Sharecropping in 1427 Florence', in K.D. Kauffman (ed.) *Advances in Agricultural Economic History*, Vol. 1, *New Frontiers in Agricultural History*, (Stamford, CT: JAI Press, 2000), pp. 177–206; R.J. Emigh, 'The Spread of Sharecropping in Tuscany: the Political Economy of Transaction Costs', *American Sociological Review*, 62 (1997) 423–42.

7 Cohen, 'Institutions and Economic Analysis'.

8 Ellis, *Peasant Economics*, pp. 146–64.

9 G. Federico, 'The "real" puzzle of sharecropping: why is it disappearing?', *Continuity and Change*, 21 (2006) 261–85.

10 As in the case of 29 per cent sharemilkers in New Zealand, who only provide their labour. See also R.W.M. Johnson, *Reforming EU Farm Policy: Lessons from New Zealand*, (Institute of Economic Affairs, 2000).

11 J. Cohen, J. and F.L. Galassi, 'Sharecropping and productivity: feudal residues in Italian agriculture, 1911', *Economic History Review*, 2nd series, 43 (1990) 646–56; *idem*, 'The Economics of Tenancy in Early Twentieth Century Southern Italy, *Economic History Review*, 2nd series, 47 (1994) 585–600; P. Hoffman, 'The Economic Theory of Sharecropping in Early Modern France', *Journal of Economic History*, 44 (1984) 309–19; *idem*, 'Sharecropping and Investment in Agriculture in Early Modern France', *Journal of Economic History*, 42 (1982) 155–9; B. Câmara, 'The Portuguese Civil Code and the colonia tenancy contracts in Madeira, 1867–1967', *Continuity and Change*, 21 (2006) 213–33; J. Carmona, 'Sharecropping and livestock specialization in France, 1830–1930', *Continuity and Change*, 21 (2006) 235–59; R. Santos, 'Risk-sharing and social differentiation of demand in land-tenancy markets in southern Portugal, seventeenth-nineteenth centuries', *Continuity and Change*, 21 (2006) 287–312.

12 E. Evans, *The Contentious Tithe: The Tithe Problem and English Agriculture, 1750–1850*, (Routledge and Kegan Paul, 1976), pp. 16–20; R.M. Townsend, *The Medieval Village Economy*, (Princeton: Princeton University Press, 1993), p. 97.

13 See below, pp. 39–40. R.A. Dodgshon, *Land and Society in Early Scotland*, (Oxford: Clarendon Press, 1981), pp. 245–53.

14 E.M. Griffiths (ed.) *William Windham's Green Book 1673–1688*, Norfolk Record Society, Vol. LXVI (2002); *idem*, 'Responses to Adversity: the changing strategies of two Norfolk landowning families, c.1665–1700' in R.W. Hoyle (ed.) *People, Landscape and Alternative Agriculture: Essays for Joan Thirsk*, Agricultural History Review Supplement 3 (2004), pp. 74–92.

15 R. Suggett, *Houses and History on the March: Radnorshire 1400–1800*, (Ceredigion: RCAHMW, 2005), pp. 181–210.

16 Dodgshon, *Land and Society in Early Scotland*, pp. 245–53; *idem*, 'Coping with Risk: Subsistence Crises in the Scottish Highlands and Islands, 1600–1800', *Rural History*, 15 (2004) 1–25; T.C. Smout, *A History of the Scottish People, 1560–1830*, (Collins, 1969), p. 131; J.E. Handley, *Scottish Farming in the Eighteenth Century*, (Faber, 1953), pp. 33–74, refers to the Highland system of steelbow as *leth-cas* or 'half-foot'. *The Oxford English Dictionary* cites several references including one that compared it to the *métayers* of France, or the half-foot tenants of the Hebrides. Sir John Sinclair described it as 'another method of occupying a farm equally barbarous in itself and adverse to improvement'.

17 R.H. Hilton, 'Why was there so little champart rent in Medieval England?', *Journal of Peasant Studies*, 17 (1990), p. 513.

18 R.H. Britnell, *Britain and Ireland 1050–1530, Economy and Society*, (Oxford: Oxford University Press, 2004), pp. 401–4.

19 F.J. Hall and P. Martyn, *Changes in Sharemilking: 1973–1993*, Technical Paper, 11 (Ministry of Agriculture, New Zealand, 1993); G. Taylor, *A Review of Sharemilking: 1972–1996*, (Rural Policy Unit, Ministry of Agriculture, New Zealand, 1996).

20 See below p. 48.

21 See below, pp. 47–51.

22 See below pp. 133, 137, 147–9.

23 See below, pp. 64–6.

24 J. Pickard, 'Shepherding in Colonial Australia', *Rural History* 19 (2008) 55–80; see below pp. 151–2.

25 See below, pp. 171–6.

26 See below, pp. 59, 87, 96–101, 108, 117.

27 See below, p. 59.

28 J.E. Thorold Rogers, *Six Centuries of Work and Wages: the History of English Labour*, (11th edn., T. Fisher Unwin, 1912), p. 279; B. Harvey, 'The leasing of the Abbot of Westminster's Demesne in the later middle ages', *Economic History Review* 2nd series, 22 (1969) 17–27; see below, pp. 21, 34–6.

29 See below, pp. 117–18.

30 See below, pp. 121–5.

31 See below, p. 34.

32 See below, pp. 48, 53–4, 56, 72–7, 81–3, 90–1, 95–6, 104, 109, 112, 116–18, 121–5.

33 See below, pp. 23, 34, 36–8.

34 Hilton, 'Champart rent', pp. 509–13.

35 See below, p. 83.

36 Overton, *Agricultural Revolution*, pp. 145–6.

Chapter 3

1   E. Kerridge, *Agrarian Problems in the Sixteenth Century and After*, (George Allen and Unwin, 1969), pp. 48–52.
2   R.H. Hilton, 'Why was there so little champart rent in Medieval England?', *Journal of Peasant Studies*, 17 (1990) 509–19.
3   J.E. Thorold Rogers, *Six Centuries of Work and Wages: The History of English Labour*, (11th edn, T. Fisher Unwin, 1912), p. 279.
4   R.V. Lennard, *Rural England 1086–1135: A Study of Social and Agrarian Conditions*, (Oxford: Clarendon Press, 1959), p. 194; J.A. Venn, *The Foundations of Agricultural Economics*, (Cambridge: Cambridge University Press, 1923), pp. 43–7.
5   G.C. Broderick, *English Land and English Landlords*, (Cassell, 1881), pp. 427–35.
6   M.M. Postan, *Essays on Medieval Agriculture and General Problems of the Medieval Economy*, (Cambridge: Cambridge University Press, 1973), pp. 134–49.
7   For example, B. Harvey, *Westminster Abbey and its Estates in the Middle Ages*, (Oxford: Clarendon Press, 1977); G.C. Homans, *English Villagers of the Thirteenth Century*, (Cambridge, Mass.: Harvard University Press, 1942).
8   B.M.S. Campbell, 'The Agrarian Problem in the Early Fourteenth Century', *Past & Present*, 188 (2005), p. 10.
9   See above, Table 2.1, p. 14.
10   Hilton's paper was given at the Flaran conference, 1985, and subsequently published in *Les Revenues de la Terre: complant, champart, métayage en Europe occidentale (IXe–CVIII siècles)*, (Auch 1987), before appearing in the *Journal of Peasant Studies* in 1990.
11   Hilton, 'Champart rent', pp. 512–13.
12   *Ibid.* p. 513.
13   A.J. Kettle, 'Agriculture 1300–1540', in G.C. Baugh (ed.) *The Victoria History of the County of Shropshire, Vol. 4, Agriculture*, (Oxford: Oxford University Press for the Institute of Historical Research, 1989), pp. 72–118.
14   *Ibid.* p. 105.
15   Hilton, 'Champart rent', p. 515.
16   *Ibid.* p. 513.
17   *Ibid.* pp. 513–14.
18   Homans, *English Villagers*, p. 202; Harvey, *Westminster Abbey*, pp. 138, 153, 320.
19   Hilton, 'Champart rent', p. 515.
20   R.H. Hilton, *The English Peasantry in the Later Middle Ages*, (Oxford: Clarendon Press, 1973), p. 43.
21   *Ibid.* p. 50.
22   Postan, *Essays on Medieval Agriculture*, pp. 134–49.
23   *Ibid.* pp. 136–7.
24   *Ibid.* p. 136, n. 65.
25   *Ibid.* pp. 140–1, n. 73.
26   *Ibid.* p. 141.
27   *Ibid.* pp. 147–9.

28 F.L. Galassi, 'Moral Hazard and Asset Specificity in the Renaissance: The Economics of Sharecropping in 1427 Florence', in K.D. Kauffman (ed.) *Advances in Agricultural Economic History*, Vol. 1, *New Frontiers in Agricultural History*, (Stamford, CT: JAI Press, 2000), pp. 177–206.

29 Lennard, *Rural England*, pp. 189–196. See above, p. 18.

30 M.M. Postan, *The Medieval Economy and Society*, (Weidenfeld and Nicolson, 1972), p. 125.

31 *Ibid.* p. 136.

32 J.A. Raftis, *Tenure and Mobility: Studies in the Social History of the Mediaeval English Village*, (Toronto: Pontifical Institute of Mediaeval Studies, 1964), pp. 44, 76, includes two references to sharecropping but offers no analysis. E. Miller and J. Hatcher, *Medieval England: Rural Society and Economic Change, 1086–1348*, (Longman, 1978) placed great emphasis on the communal nature of village life and the extent of subletting and inter-peasant leasing, leaving the tenurial pattern in a constant state of flux. They refer to 'The existence of men eager to take up plots of customary land, even on a sharecropping basis, was assumed in the customs of Bradford on Tone', p. 143, but there is no further comment. See also, J. Hatcher, *Rural Economy and Society in the Duchy of Cornwall, 1300–1500*, (Cambridge: Cambridge University Press, 1970) for emphasis on the extent of subletting in the county, pp. 16, 68, 139, 219, 231–5, 237–8, 252.

33 Campbell, 'The Agrarian Problem', p. 10.

34 *Ibid.* p. 47.

35 *Ibid.* p. 55.

36 *Ibid.* p. 54, n. 161.

37 *Ibid.* p. 57.

38 The full extent of subletting is revealed by M.K. McIntosh, 'Land, Tenure and Population in the Royal Manor of Havering, Essex, 1251–1352', *Economic History Review*, 2nd series, 33 (1980) 17–31. Successive surveys of 1251 and 1352/3 record a class of 'undersettles' holding very small parcels of land from other men, giving rise to chains of subletting with those at the end paying the most exorbitant rent. Short of squatting there was no other way the poorest could gain a foothold on the land; the chains show that the tenants were the rack renters rather than the landlords.

39 A.V. Chayanov, *The Theory of Peasant Farming*, ed. D. Thorner, B. Kerbaly and R.E.F. Smith, (Homewood, Ill.: American Economic Association, 1966), pp. 53–69.

40 R.M. Smith (ed.) *Land, Kinship and Life Cycle*, (Cambridge: Cambridge University Press, 1984).

41 J. Ravensdale, 'The transfer of customary land on a Cambridgeshire manor [Cottenham] in the fourteenth century', in Smith, *Land, Kinship and Life-Cycle*, p. 218.

42 S. Ogilvie, '"Whatever is, is right"? Economic institutions in pre-industrial Europe', *Economic History Review*, 2nd series, 60 (2007) 649–84.

43 R.M. Smith, 'Families and their land in an area of partible inheritance: Redgrave, Suffolk, 1260–1320', in Smith, *Land, Kinship and Life-Cycle*, pp. 135–95.

44 R.M. Smith, 'Some issues concerning families and their property in rural England, 1250–1800', in Smith, *Land, Kinship and Life-Cycle*, pp. 1–86;

E. Clark, 'Some Aspects of Social Security in Medieval England', *Journal of Family History*, 7 (1982).

45 Smith, 'Some issues concerning families'; C. Howell, *Land, Family and Inheritance in Transition: Kibworth Harcourt*, (Cambridge: Cambridge University Press, 1983), pp. 252–3.

46 Z. Razi, 'Family, Land and the Village Community in Later Medieval England', *Past & Present*, 93 (1981) p. 10.

47 *Ibid.* pp. 32–3.

48 Lennard, *Rural England*, p. 194.

49 Thorold Rogers, *Six Centuries of Work and Wages*, p. 279.

50 *Ibid.* p. 279. C.C. Dyer, 'Farming practice and techniques: the West Midlands', in E. Miller (ed.) *The Agrarian History of England and Wales, Vol. III, 1348–1500*, (Cambridge: Cambridge University Press, 1991), p. 235, describes the huge scale of the enterprises in the West Midlands with 2,500 sheep leased with pastures in Blockley, Glos. He also explains that the normal method of managing cows was to lease out the lactage of each animal at 5s per annum. R. Suggett says that while graziers entered into sharecropping agreements in Radnorshire, they hired flocks and herds of cattle in Cardiganshire and west Wales. (pers. comm.).

51 Thorold Rogers, *Six Centuries of Work and Wages*, p. 279.

52 B. Harvey, 'The leasing of the Abbot of Westminster's Demesne in the later middle ages', *Economic History Review*, 2nd series, 22 (1969) 17–27.

53 *Ibid.* p. 21. E. Miller, 'Tenant farming and tenant farmers: the southern counties', in E. Miller (ed.) *The Agrarian History of England and Wales, Vol. III, 1348–1500*, (Cambridge: Cambridge University Press, 1991), p. 713, questions how leaseholders acquired the capital to take up demesne lands. The terms on which John Gervays held Enford in the 1430s came near to a sharecropping agreement. He started off with so much land sown and a stock of pigs and poultry that enabled him to pay his rents in liveries of grain, pigs, poultry, geese and capons. In other words, he secured a long term loan to get started. Postan, in his introduction to *The Book of William Morton: Almoner of Peterborough Monastery, 1448–1467*, ed. P.I. King, Northants Records Society, 16 (1954), pp. xxxv–xxxvi describes the almoner 'lending cows and calves, selling hay and leasing grazing'. The loan of cows and calves, he suggested, was a cover up for a little profitable money lending. Annual payments for the use of borrowed cows was not considered usury. Loans on cows used to legitimize petty borrowing on interest. When the almoner lent a cow for a year or two, there was no reason to regard it other than what it purports to be, but when the loan was for five years, or for a ten year old cow for three years at 3s per annum, with an understanding to pay a particular sum if the cow could not be returned, the impression of a disguised loan is inescapable.

54 Harvey, 'The Leasing of the Abbot of Westminster's Demesne', p. 23.

55 Harvey, *Westminster Abbey*, pp. 138, 153, 319–20. Besides the references quoted by Hilton, Harvey mentions a more detailed agreement at Yeoveney, in 1308–9, where £115 was spent on seed corn, fodder and liveries for the demesne workers, the sum being the price of half the crop of wheat and maslin.

56 R.H. Britnell, *Britain and Ireland, 1050–1530: Economy and Society*, (Oxford: Oxford University Press, 2004), pp. 401–4.

57 R.H. Britnell, 'The Pastons and their Norfolk', *Agricultural History Review*, 36 (1988) 132–44; C. Richmond, 'Landlord and Tenant: the Paston Evidence', in J. Kermode (ed.) *Enterprise and Individuals in Fifteenth Century England*, (Stroud: Alan Sutton, 1996), pp. 25–42, on the difficulties the family encountered collecting rents, and the payment of rent in meat and grain.

58 Norfolk Record Office (hereafter NRO), PHI 532 578 X 2.

59 Venn, *Foundations of Agricultural Economics*, pp. 36–7, notes that in 1845 the Master of Corpus Christi College publicized the success of corn rents adopted in lieu of granting further reductions in money rents. Both parties benefited: the tenant was relieved when prices were low, and proportionally benefited when prices were high.

60 B.M.S. Campbell, 'Agricultural Progress in Medieval England: Some Evidence from Eastern Norfolk', *Economic History Review*, 2nd series, 36 (1983) 26–46; R.H. Britnell, 'The occupation of the land: Eastern England', in E. Miller (ed.) *The Agrarian History of England and Wales, Vol. III, 1348–1500*, (Cambridge: Cambridge University Press, 1991), pp. 53–66; *idem*, 'Farming practice and techniques: Eastern England' in Miller (ed.) *Agrarian History of England and Wales, Vol. III*, pp. 194–209.

61 Britnell, 'The Pastons and their Norfolk', p. 134.

62 Britnell, *Britain and Ireland*, p. 402.

63 NRO, LEST/BK7; NRO, LEST/IC58; NRO, LEST/OC1.

64 C.E. Moreton, *The Townshends and their world: Gentry, Law, and Land in Norfolk c.1450–1551*, (Oxford: Clarendon Press, 1992), pp. 146–7; C. Richmond, *John Hopton: a Fifteenth Century Suffolk Gentleman*, (Cambridge, Cambridge University Press, 1981), pp. 31–99. His daughter Eleanor married Roger Townshend. As at Raynham, information of demesne farming is thwarted by the nature of bailiff accounts, which excluded information on livestock and arable farming. However, the accounts of the bailiff of husbandry and the sheep reeve, show that flexible arrangements were common. In 1490s when Thomasina let the demesne at Cockfield Hall with her dairy, she took a good part of the rent in kind. The two principal tenants paid half their rent in kind; one supplied the household with butter milk, cheese, chickens, geese, oats and malt; carted wood and paid some cash. The other delivered fish, geese, faggotts, ground corn and pastured 3 of Thomasina's cows.

65 R.W. Hoyle, 'Agrarian agitation in mid-sixteenth-century Norfolk: A petition of 1553', *Historical Journal*, 44 (2001), p. 237.

66 Britnell, *Britain and Ireland*, p. 401.

67 The proportion of the crop paid by the tenant is an interesting aspect of the system. In twentieth-century New Zealand a sharemilker could enter into different types of arrangement, starting with a 29 per cent agreement whereby he supplied little more than his labour; 39 per cent when he provided his equipment, a tractor and trailer; 50 per cent when he provided the herd. In a country like New Zealand where wages were high, and the contract invariably included family labour, the 29 per cent agreements were rare; most entered at 39 per cent, quickly moving up to 50 per cent where the substantial returns were made.

68 R.A. Dodgshon, *Land and Society in Early Scotland*, (Oxford: Clarendon Press, 1981), pp. 245–53.

69 Steelbow is compared to German tenure where stock was leased with the land, with rent often being paid in kind, like the *métayers* of France.
70 As with corn rents in medieval Norfolk, steelbow was often associated with the collection of tithes. See above, pp. 19, 37.
71 Dodgshon, *Land and Society in Early Scotland*, p. 245.
72 *Ibid.* p. 251.
73 A. Smith, *An Inquiry into the Nature and Causes of the Wealth of Nations*, (1776), Book 3, Part 2.
74 T.C. Smout, *A History of the Scottish People, 1560–1830*, (Collins, 1969), p. 131; J.E. Handley, *Scottish Farming in the Eighteenth Century*, (Faber, 1953), pp. 33–74, refers to the Highland system of steelbow as *leth-cas* or 'half-foot'.
75 R.A. Dodgshon, 'Coping with Risk: Subsistence Crises in the Scottish Highlands and Islands, 1600–1800', *Rural History*, 15 (2004) 1–25.
76 With no constraints on subdivision, subletting through middlemen became endemic in Ireland in the 1600s and was not resolved until after the Famine. K. O'Neill, *Family and Farm in Pre-Famine Ireland: The Parish of Killashandra*, (London: University of Wisconsin Press, 1984), pp. 33–5; E.D. Steele, *Irish Land and British Politics: Tenant Right and Nationality, 1865–1870*, (Cambridge: Cambridge University Press, 1974), pp. 7–8.
77 Smout, *A History of the Scottish People*, pp. 111–18.
78 From a discussion with the French historian, Francis Brumont. See above, p. 2.

Chapter 4

1 R. Suggett, *Houses and History on the March: Radnorshire 1400–1800*, (Ceredigion: RCAHMW, 2005), pp. 181–210.
2 See above, Table 2.1, p. 14.
3 P. Edwards, *Farming: Sources for Local Historians*, (Batsford, 1991), pp. 32–47.
4 E. Kerridge, *Agrarian Problems in the Sixteenth Century and After*, (George Allen and Unwin, 1969), pp. 48–52.
5 C.J. Harrison, 'Elizabethan Village Surveys: a comment', *Agricultural History Review*, 27 (1979) 82–9.
6 Kerridge, *Agrarian Problems*, p. 52.
7 G.E. Fussell (ed.) *Robert Loder's Farm Accounts 1610–1620*, Camden Society, 3rd series, Vol. 53 (1936), pp. xxiv, 1, 19, 34.
8 *Ibid.* pp. 72, 90, 122.
9 E. Kerridge, *The Agricultural Revolution*, (George Allen and Unwin, 1967), p. 194.
10 J.H. Bettey, 'The Floated Water-Meadows of Wessex: a Triumph of English Agriculture', in H. Cook and T. Williamson (eds) *Water Meadows: History, Ecology and Conservation*, (Macclesfield: Windgather Press, 2007), p. 12.
11 Suggett, *Houses and History on the March*, pp. 181–210. He refers to share-farming as sharecropping, even though it only applied to livestock.
12 F.V. Emery, 'The Farming Regions of Wales', in J. Thirsk (ed.) *The Agrarian History of England and Wales, Vol. IV, 1500–1640*, (Cambridge: Cambridge University Press, 1967), pp. 137–9, describes lean cattle from Herefordshire,

Shropshire and Staffordshire and the Midlands grazing the summer pastures of the central uplands. Emery provides a single example of the tenant of Wynns, Gwdir in the 1570s being given 8 cows and paying as part of his rent 'half the cheese or twenty shillings'.

13   Suggett traced more than 40 agreements which offered precise combinations of sheep and cattle, and occasional references to 'sharecropping' with horses in west Radnorshire.

14   Suggett, *Houses and History on the March*, p. 183.

15   *Ibid.* pp. 183–7.

16   See below, pp. 156–7.

17   Lichfield Joint Record Office (hereafter LJRO), B/C/11 7 March 1638/9, John Beresford the Younger of Wellington; B/C/11, 30 April 1635, John Beresford of Wellington; B/C/11, 30 July 1606, Wm. Povas of Whitchurch; B/C/11, 2 October 1677, Tho. Parbott of Ellcom; B/C/11, 17 August 1608, Humphrey Brott of Child's Ercall; B/C/11, 6 October 1632, John Morgan of Great Bolas; B/C/11, 10 July 1639, Edward Capper of Baschurch.

18   LJRO, B/C/11, 17 June 1606, John Charlton, Ercall Magna; B/C/11, 3 October 1634, Rob. Eccles, Drayton in the Hales; L.J.R.O., B/C/11, 18 June 1600, Wm. Chidlowe of Burleton Loppington. We are grateful to Peter Edwards for these examples from his transcripts of Shropshire inventories.

19   M. Overton, 'Agricultural change in Norfolk and Suffolk, 1580–1740', (Unpublished PhD thesis, University of Cambridge, 1981); the inventories are from the Norwich Consistory Court and were transcribed in the 1970s. They are used to best effect in B.M.S. Campbell and M. Overton, 'A new perspective on medieval and early modern agriculture: six centuries of Norfolk farming, c.1250–c.1850', *Past & Present*, 141 (1993) 38–105.

20   Norfolk Record Office, MFX/INV/3/001; INV/39/311; INV/34/26.

21   The Cornwall and Kent inventories were collected as part of a project funded by the Leverhulme Trust and are described in M. Overton, J. Whittle, D. Dean and A. Hann, *Production and Consumption in English Households, 1600–1750*, (Routledge, 2004), pp. 28–31.

22   Cornwall Record Office (Hereafter CRO) Inventory C414.

23   CRO Inventory D515.

24   CRO Inventory T283.

25   Other farming categories included: horses, mares and colts mentioned in 14 inventories, hay in 5, wool in 3, goats in 3 and bees in 2.

26   G.B. Worgan, *General View of the Agriculture of the County of Cornwall*, (1811). See below, pp. 124–5.

27   P. Edwards, 'Agriculture 1540–1750' in G.C. Baugh (ed.) *The Victoria History of the County of Shropshire, Vol. 4, Agriculture*, (Oxford: Oxford University Press for the Institute of Historical Research, 1989), p. 151; BL Add. MS 33509. fol. 74; A. Speed, *Adam out of Eden*, (1659). See also, C.F. Foster, *Seven Households: Life in Cheshire and Lancashire, 1582–1774*, (Northwich: Arley Hall Press, 2002), pp. 91–5.

28   Centre for Kentish Studies (hereafter CKS) 11.61.2 and 11.23.2.

29   CKS 11.11.129. See also CKS 11.28.124 and 10.51.193.

30   Berkshire Record Office, D/A1/46/112 and BRO D/A1/92/133.

31   Since sharefarming was not systematically recorded in Norfolk we cannot express its incidence as a proportion of total inventories. For Cornwall the

seventeenth-century sample consists of 2041 inventories in which farming of some sort, however small-scale, was taking place and for Kent 2127.

32 National Archives E 134, 4 Wm & Mary, Mich 43, Luke Leake v. Samuel Warner 1692, depositions of John Smyth, Great Cornard, husbandman, & Daniel Cook of Sudbury, Gt., 'Daniel Cook, gt deposed that in 1688–90, Sam Warner held 14 acres of arable', sharing the lands with Stephen Carter 'according to the usuall way or manner of halving ... Warner to find the land, barn and pay taxes & Carter to plough, sow and cut the corn. Sam Warner deposed that Carter had told him he received half the crop'.

33 C.W. Chalklin, *Seventeenth-century Kent*, (Longmans, 1965), p. 63.

34 J.A. Venn, *Foundations of Agricultural Economics*, (Cambridge: Cambridge University Press, 1923), pp. 43–8. No date is given for the agreement, but it is c.1650.

35 *Ibid.* pp. 47–8.

36 W.O. Massingberd, *History of the Parish of Ormsby-cum-Ketsby*, (Lincoln, 1893), pp. 303–4; B.A. Holderness, 'The Agricultural Activities of the Massingberds of South Ormsby, Lincolnshire 1638–c.1750', *Midland History*, 1 (1972) 15–25.

37 Lincolnshire Record Office (hereafter LRO) MM/VI/1/5.

38 B.A. Holderness, 'Credit in English Rural Society before the Nineteenth Century with special reference to the period 1650–1720', *Agricultural History Review*, 24 (1976) 97–109.

39 *Ibid.* p. 103.

40 In the 19th century, the Lincolnshire Custom was used as the basis of the new law of Tenant Right, 1875 and 1883, which recognised the inputs made by the departing tenants, the idea being to prevent the deterioration of farms towards the end of their tenure. J.A. Perkins, 'Tenure, Tenant Right and Agricultural Progress in Lindsey, 1780–1850', *Agricultural History Review*, 23 (1975) 1–22.

41 Foster, *Seven Households*, pp. 91–5.

42 *Ibid.* p. 92.

43 Verney MSS BL M636/1; J. Broad, *Transforming English Rural Society: the Verneys and the Claydons, 1600–1820*, (Cambridge: Cambridge University Press, 2004), p. 65.

44 D. Woodward (ed.) *The Farming and Memoranda Books of Henry Best of Elmswell, 1642*, Records of Social and Economic History, new series, 8 (1984), p. 179.

45 LRO, DDPt/19 (agisters accounts 1612–30).

46 M. Spufford, *Contrasting Communities: English Villagers in the Sixteenth and Seventeenth Centuries*, (Cambridge: Cambridge University Press, 1974), p. 140.

47 Fussell, *Robert Loder*, p. 1.

48 K. Wrightson, *Earthly Necessities: Economic Lives in Early Modern Britain*, (Yale University Press, 2000), pp. 159–71; G. Clark, 'The Price History of English Agriculture, 1209–1914', *Research in Economic History*, 22 (2004) 41–124.

49 S. Hartlib, *His Legacie or An Enlargement of the Discourse of Husbandry used in Brabant and Flanders*, (1652), p. 10.

50 P. Hoffman, *Growth in a Traditional Society: the French Countryside 1450–1815*, (Princeton: Princeton University Press, 1996), pp. 65–9; *idem,*

'The Economic Theory of Sharecropping in Early Modern France', *Journal of Economic History*, 44 (1984) 309–19.

51 O. de Serres, *Le Théâtre d'Agriculture*, (Paris: 1600), pp. 45–50.
52 R. Santos, 'Risk-sharing and social differentiation of demand in land tenancy markets in southern Portugal, seventeenth-nineteenth centuries', *Continuity and Change*, 21 (2006) 287–312.

Chapter 5

1 E.M. Griffiths, 'Sir Henry Hobart: a new hero of Norfolk agriculture?', *Agricultural History Review*, 46 (1998) 15–34; *idem*, 'Responses to adversity: the changing strategies of two Norfolk landowning families' in R.W. Hoyle (ed.) *People, Landscape and Alternative Agriculture: Essays for Joan Thirsk*, Agricultural History Review Supplement 3 (2004), pp. 74–92; *idem, William Windham's Green Book 1673–1688*, Norfolk Record Society, Vol. LXVI (2002). Background and context to Norfolk farming in the early modern period can be found in J. Whittle, *The Development of Agrarian Capitalism: Land and Labour in Norfolk 1440–1580*, (Oxford: Clarendon Press, 2000); J. Thirsk 'The farming regions of England', in *idem* (ed.) *The Agrarian History of England and Wales, Vol. IV, 1500–1640*, (Cambridge: Cambridge University Press), pp. 40–8; B.A. Holderness, 'East Anglia and the Fens: Norfolk, Suffolk, Cambridgeshire, Ely, Huntingdonshire, Essex and the Lincolnshire Fens', in J. Thirsk (ed.) *The Agrarian History of England and Wales, Vol. V, 1640–1750. I. Regional farming systems*, (Cambridge: Cambridge University Press, 1984), pp. 197–238; A. Simpson, *The Wealth of the Gentry 1540–1660: East Anglian Studies*, (Cambridge: Cambridge University Press, 1963); B.M.S. Campbell and M. Overton, 'A new perspective on medieval and early modern agriculture: six centuries of Norfolk farming, c.1250–c.1850', *Past & Present*, 141 (1993) 38–105; M. Overton and B.M.S. Campbell, 'Norfolk livestock farming 1250–1740: a comparative study of manorial accounts and probate inventories', *Journal of Historical Geography*, 18 (1992) 377–96.
2 Griffiths, 'Sir Henry Hobart', pp. 15–34; see below, pp. 000–000 where the evidence of farming to halves at Langley appears in 1718, and Robert Britiffe's references to undertenants.
3 Norfolk Record Office (hereafter NRO), LEST/P10.
4 NRO, LEST/Q38. By 1616, Sir Hamon estimated 'the revenues of my lands' at £1,247. In 1609 he purchased a manor at Heacham, but still paid annuities of £300 a year. In 1620 Alice inherited her father's estate at Sedgeford adding a further £473 to their income. This compares to £2,500 enjoyed by his first cousin, Sir John Hobart of Blickling in 1625.
5 Farmalls list the tenants paying farm rents, as opposed to rentals which list copyhold and freehold rents. They were drawn up at both Raynham and Hunstanton, although the word 'farmall' does not appear in the Oxford English Dictionary.
6 Maps: Hunstanton, NRO, LEST/A01; Ringstead NRO, LEST/OB5, 6; Sedgeford, NRO, LEST/OC1; Heacham, NRO, LEST/OC2.
7 NRO, LEST/R8.

8   See above, Table 2.1, p. 14. A foldcourse was the exclusive right to erect a sheep fold on the fallow and sometimes also over areas of permanent pasture. Ownership of foldcourses was a monopoly of manorial lords, and tenants had very limited rights to graze their animals on the fallow. See, K.J. Allison, 'The sheep-corn husbandry of Norfolk in the sixteenth and seventeenth centuries', *Agricultural History Review*, 5 (1957) 12–30 and A. Simpson, 'The East Anglian foldcourse: some queries', *Agricultural History Review*, 6 (1958) 87–96.

9   NRO, LEST/Q38; NRO, LEST/P6.

10 NRO, LEST/BK7, LEST/1C58; LEST/89, LEST/IB90.

11 NRO, LEST/BK3.

12 NRO, LEST/P6.

13 See above, pp. 45–6.

14 NRO, LEST/KA6: small book of expenses on the marshes, 1633–1640; NRO, LEST/KA9: ploughing and sowing accounts, 1643–1645; NRO, LEST KA10: notes by Sir Hamon and Sir Nicholas; NRO, LEST KA24: Book of Surveys, 1640–1643.

15 NRO, LEST/KA6.

16 C. Oestmann, *Lordship and Community: The Lestrange Family and the Village of Hunstanton in the First Half of the Sixteenth Century*, (Woodbridge: The Boydell Press, 1994), p. 111, makes no mention of farming to halves, but emphasizes the sharing and lending of carts, ploughs and equipment.

17 NRO, LEST/P7. Expenditure on improvements continued throughout the Civil Wars even though the Le Stranges were involved in the defence of King's Lynn and lost their sheep at the hands of the Parliamentary forces in 1643. R.W. Ketton-Cremer, *Norfolk in the Civil War*, (Faber & Faber, 1969), pp. 206–18.

18 NRO. LEST, KA9. See the Frontispiece.

19 NRO, LEST, KA24.

20 Sir Nicholas Le Strange's four missing notebooks: NRO, LEST, KA4, KA5, KA7, KA8.

21 NRO, LEST, P7.

22 L. Campbell, *Sir Roger Townshend and his Family: A Study of Gentry Life in Early Seventeenth Century Norfolk*, (Norwich: Centre of East Anglian Studies, 1990).

23 Townshend MSS (hereafter TM), RAS/F2/9.

24 TM, RAS/F2/5 RAS/F2/12 Langham and Morston 1632, worth £200; Stiffkey 1646, worth £600.

25 TM, RAS/F2/10 Stibbard and Ryburgh 1623. C.E. Moreton, *The Townshends and their World: Gentry, Law and Land in Norfolk c.1450–1551*, (Oxford: Clarendon Press, 1992), pp. 146–7.

26 The estate at *c.* £2000 a year was comparable to that of Sir John Hobart of Blickling.

27 TM, RAS/F2/6; F2/9.

28 TM, RAS/F2/5, in 1615, Lady Jane also purchased lands in South Creake worth £100 a year.

29 F. Blomefield, *An Essay Towards a Topographical History of the County of Norfolk*, (11 vols, 1805–10), Vol. 7, pp. 152–8. The *History* was first published 1739–75.

30 TM, RAS/F2/6.
31 TM, RAS/F2/5.
32 TM, RAS/F2/9.
33 TM, RAS/2/6.
34 TM, RAS/F2/9.
35 TM, RAS/F2/9.
36 TM, RAS/F2/10.
37 TM, RAS/F2/11.
38 TM, RAS/F2/11.
39 TM, RAS/F2/12; RAS/F3/1.
40 TM, RAS/F2/12.
41 TM, RAS/F2/12.
42 BL Add. MS 41308. Kipton is a few miles from Raynham. Edward Symonds, accountant for the trustees of Sir Roger Townshend, 1636–1643, was succeeded by his brother William on his death in 1644.
43 TM, RAS/F2/10.
44 TM, RAS/F2/12.
45 TM, RAS/F3/4.
46 See above, pp. 27–30.
47 TM, RAS/F3/1.
48 TM, RAS/F3/4.
49 TM, RAS/F3/1.
50 TM, RAS/F3/4. A note, from the early 1660s, recording the 'lands held by principal tenants' showed Peter Stringer, renting lands worth £291 a year, John Robotham £213, Richard King of South Creake £264 and John Raby of Shereford, £100. RAD/C1.
51 TM, RAS/F3/4.
52 TM, RAD/A5.
53 TM, RAD/A5.
54 TM, RAD/A4.
55 TM, RAD/A4.
56 E. Kerridge, *The Farmers of Old England*, (George Allen and Unwin, 1973), pp. 103, 147–8.
57 E.M. Griffiths, 'The management of two East Norfolk Estates in the Seventeenth Century: Blickling and Felbrigg', (Unpublished PhD thesis, University of East Anglia, 1987), pp. 235–43. For example, in 1635, for Bush and Hill Close on the edge of Felbrigg Park, Windham directed that the former should be broken up and tilled the first five years of the term, and the latter laid to pasture, while in the last five years the reverse would apply. And when the closes were tilled, oats were to be sown only once. The policy extended to every part of the estate.
58 Griffiths, 'The management of two East Norfolk Estates', pp. 209–22. Thomas Windham inherited Felbrigg from his father Sir John Wyndham of Orchard, Somerset in 1616. Sir John had acquired the reversion of the estate from the notorious Roger in 1598 and gained possession in 1608.
59 John Thompson of Felmingham who leased the Sustead Meadows between 1611–16.
60 Sir Robert Bell's grandfather, another Sir Robert, came to Stow Bardolph in the mid sixteenth century and married the daughter of Edmund Beaupre

of Outwell. His elder daughter Mary married Sir Nicholas Le Strange (Sir Hamon's father) while the younger, Dorothy married Sir Henry Hobart. Sir Robert Bell, Sir Hamon Le Strange and Sir John Hobart were first cousins. All three families were involved in various drainage schemes. Sir John acquired Sir Robert's estate at Stow Bardolph in repayment of debt. The sale was not completed until the 1650s, when the Hobarts sold the estate, but in the meantime Sir John managed the property as best he could.

61   R. Allen, *Enclosure and the Yeoman: The Agricultural Development of the South Midlands, 1450–1850*, (Oxford: Clarendon Press, 1992).

62   Overton, *Agricultural Revolution*, pp. 133–92.

63   Griffiths, *William Windham's Green Book*, pp. 27–34; *idem*, 'Responses to adversity'.

64   NRO, WKC 5/152, 400 X.

65   Sir Joseph Ashe (1618–1686) of Cambridge Park, Twickenham, was the third son of James Ashe, a London clothier with west country connections. Having built up a considerable fortune as a cloth trader, Joseph acquired estates at Downton, Wiltshire and Wawne, East Yorkshire in the 1650s. He received a baronetcy in 1660, and served as M.P. for Downton from 1662–1681. He continued to live at Twickenham, but invested heavily in agricultural improvement, including the creation of water meadows. See J.H. Bettey, *Wiltshire Farming in the Seventeenth Century*, Wiltshire Record Society, Vol. 57 (2005), pp. 236–75; *idem*, 'The Development of water meadows on the Salisbury Avon, 1665–1690', *Agricultural History Review*, 51 (2003) 163–72; M. Carrick, 'Lords of the Manor of Wawne: the Ashe, Windham and Smith Families 1651–1950', *East Yorkshire Local History Society*, 47 (1992) 15–21; Sir Joseph Ashe may have become acquainted with the Windhams of Felbrigg, through their first cousins, the Windhams of north Wiltshire.

66   R.W. Ketton-Cremer, *Felbrigg: The Story of a House*, (Ipswich: The Boydell Press, 1962), pp. 44–77.

67   T. Browne, *The Garden of Cyrus*, (1658); S. Hartlib, *His Legacie or An Enlargement of the discourse of Husbandry used in Brabant and Flanders*, (1652).

68   NRO, WKC, 7/15, 404 X 1.

69   NRO, WKC, 5/442, 464 X 4.

70   NRO, WKC, 5/151, 400 X 5; NRO, WKC, 5/152, 400 X.

71   Griffiths, *William Windham's Green Book*, p. 272.

72   NRO, WKC, 7/154, 404 X 8.

73   Griffiths, *William Windham's Green Book*, pp. 171–4.

74   NRO, WKC, 5/151, 400 X 5; NRO, WKC, 5/154 X 6.

75   Griffiths, *William Windham's Green Book*, pp. 23, 150–2, 277.

76   *Ibid.* pp. 150–3.

77   Edmund was the son of Edmund Britiffe, who had advised Thomas Windham on his grazing agreements in the 1620s. He may well have drawn up the lease for Rush Close. See below, p. 70.

78   Griffiths, *William Windham's Green Book*, pp. 283–5.

79   NRO, WKC, 5/158, 400 X 6.

80   Griffiths, *William Windham's Green Book*, pp. 266–7.

81 *Ibid.* pp. 16, 213–4, 249, 263.

82 *Ibid.* pp. 206–8, 247, 249–50.

83 NRO, WKC, 5/158 X 6; E.M. Griffiths, 'The management of two East Norfolk Estates', pp. 388–97.

84 O. de Serres, *Le Théâtre d'Agriculture*, (Paris: 1600), pp. 45–50 reminds us that sharefarming was a useful way of collecting rent.

85 The changes in the estate rental were due, firstly to the death of William Windham's mother in 1679, and secondly his own death in 1689, when Katherine took the farms in Suffolk and Essex worth £600 a year.

86 Griffiths, *William Windham's Green Book*, pp. 167–9.

87 Katherine Windham experienced her worst debacle in 1694, when Jeremiah Prance left Middleton Hall, Essex, with debts of £555. NRO, WKC 5/158, 400 X 6.

88 In 1678, Windham had rescinded an abatement of £10 from John Kingsberry of Middleton Hall 'because he had spent too much time in the wheelwright's company'. In 1682, the relationship reached crisis point over Kingsberry's mistreatment of young trees, so he 'resolved to run the hazard of getting a new tenant', with disastrous consequences.

89 R.H. Hilton, 'Why was there so little champart rent in Medieval England?', *Journal of Peasant Studies*, 17 (1990), p. 517.

90 Griffiths, 'Responses to adversity', 85–90. The section on Blickling is drawn largely from this article.

91 Henry Gallant was one of John Brewster's men on the ground.

92 S. Wade Martins and T. Williamson, 'The Development of the Lease and its Role in Agricultural Improvement in East Anglia, 1660–1870', *Agricultural History Review*, 46 (1998) 129–30.

93 NRO, NRS 16338 32C2.

94 Griffiths, 'The management of two East Norfolk Estates', pp. 410, 486.

95 The properties at Horsham St. Faith's, Hevingham and Saxthorpe were sold on the death of Sir John Hobart 3rd Bt. to repay debts. Sir Henry Hobart's eldest daughter, Henrietta received the rents for Langley, intermittently until 1718, when it was sold to the Berneys.

96 The Britiffes acquired these lands from the Bacons and the Pastons.

97 NRO, GTN 425.

98 J.M. Rosenheim, *The Townshends of Raynham: Nobility in Transition in Restoration and Early Hanoverian England*, (Middletown, Ct.: Wesleyan University Press, 1989), pp. 144, 226; *idem, The Emergence of a Ruling Order: English Landed Society 1650–1750*, (Longman, 1998); J.H. Plumb, *Sir Robert Walpole. [1], The Making of a Statesman*, (Cressett Press, 1956), pp. 310, 361; Norfolk Lists, (Norwich: 1837); A. Campling (ed.) *East Anglian Pedigrees*, Norfolk Record Society, Vol. 13 (1943), pp. 240–3.

99 Plumb, *Sir Robert Walpole*, p. 310.

100 Judith Britiffe married Sir John Hobart 5th Bt. in 1717, NRO, NRS, 15832 31F6; Elizabeth Britiffe married Sir William Morden Harbord, 1st Bt in 1732, NRO, Trafford, 234 88 x 1.

101 NRO, PET 443 98 X 3. Westwick lies close to Dilham Hall, which was broken up and let to halves in the 1650s, see above, p. 72.

102 NRO, NRS, 16301 32B5.

103 G.A. Chinnery, *A Handlist of the Cholmondeley (Houghton) Manuscripts: Sir Robert Walpole's Archive*, (Cambridge: Cambridge University Press, 1953).

104 Leasing cow grasses is still a common practice in Norfolk, stockmen hiring grazing from farmers for a specified number of cattle from May to November. The farmer is responsible for fertilizer, maintaining hedges, gates, watercourses and so on.

105 NRO, LEST/KA11; Sir Nicholas Le Strange, 4th Bt. 1661–1724.

106 Sir Nicholas Le Strange, 1st Bt. 1604–1655.

107 C. Oestmann, *Lordship and Community, The Lestrange Family and the Village of Hunstanton, Norfolk in the First Half of the Sixteenth Century*, (Woodbridge: The Boydell Press, 1994).

108 Rosenheim, *Townshends of Raynham*, pp. 64–194.

109 TM, RAS/A1/6.

110 The reorganization of land in the park at Raynham was associated with the completion of the New Hall started by Roger Townshend in the 1620s and completed by his son in the 1660s.

111 Letters from 1665/6 include references to the division of farms at East Raynham – 'an agreement made by the farmers of such lands as they intend to have of the new inclosures'; it shows five tenants sharing 505 acres divided into 39 closes: BL Add. Mss 41655.

112 Felton's correspondence 1679–1682, TM, RAS/A1/3; BL Add. Mss 41655.

113 Evidently, sales of dairy produce were slow. A letter to Lady Townshend in 1681 records that 'people eat little butter and cheese'. Townshend clearly respected his female dairy farmers, in 1683, he loaned Martha Clark of South Raynham £214 on the security of her 8 cows, 7 bullocks, 5 horses, 3 mares 2 colts and 12 swine; he also allowed Widow Cooke to continue in her substantial holding following her husband's death in 1682.

114 TM, RAS/A1/3.

115 TM, RAS/A1/3.

116 TM, RAS/G1/9.

117 The Raynham Household Acct. 1687, records wages of £5 paid to the cowman and £2.15.0 to Jane Gyles, the dairy maid. They sold butter and cheese worth £7.7.4, and sent pots of butter to London for Lord T. In 1688 her husband E. Giles leased 'ye dayrey of cows at Raynham Hall' for £50 – rent payable at Xmas – 'but the summer being drye abated by order of the executors: TM, RAS/A2/1-2.

118 TM, RAS/A1/5; RAS/A1/7; BL Add. Mss 41655. William Windham also forbade subletting at Felbrigg.

119 TM, RAS/A1/6.

120 Cow grasses continue to appear until 1712.

121 BL Add. MSS. 41656/158.

122 TM, RAS/A1/5. The bill of sale, often a feature of these agreements, was the security for the stock, in effect a mortgage.

123 TM, RAS/A1/6; RAS/A2/1-8.

124 TM, RAS/A2/1-8.

125 In 1659, Felton described how Stringer, who leased part of the park and foldcourse, had a plan to dam the river to create a water supply for the mill which he was to set up: BL Add. MSS. 41655/121.

126 Rosenheim, *Townshends of Raynham*, p. 13; A.H. Smith, G.M. Baker and R.W. Kenny (eds) *The Papers of Nathaniel Bacon of Stiffkey, Vol. 1, 1556–1577*, Norfolk Record Society, Vol. 46 (1979); *The Papers of Nathaniel Bacon of Stiffkey, Vol. 2, 1578–1585*, Norfolk Record Society, Vol. 49 (1983).
127 TM, RAS/A1/5.
128 TM, RAS/A1/5.
129 TM, BL Add. Mss 41655/160.
130 TM, RAS/A1/5.
131 A pightle is a small enclosure often attached to a tenement and leased to a cottager or smallholder.
132 TM, RAS/A1/3.
133 BL Add. MSS. 41655/166; Teasdale also leased the Morston Foldcourse.
134 'I turned Smith away for a knave', HT's Memo and Account Book, TM, RAS/G1/9.
135 1684 Farmall, TM, RAS/A1/6.
136 TM, RAS/A1/6.
137 T. Ward's correspondence, TM, RAS/A1/6, and subsequent accounts RAS/A2/1-8.
138 T. Ward's correspondence 1686–87, TM, RAS/A1/6.
139 H. Parson's lease 1686, TM, RAS/A1/7.
140 TM, RAS/A1/7; RAS/A2/1-8; RAS/A3/1-8.
141 Ward in 1706 complained, 'at Stiffkey I did not get a penny, I fear my business wil not be better at Langham ... Money is scarce than even I remember it ... I cannot promise any money until I have it, I never met with such trouble and disupoyntment': TM, RAS/A1/6; RAS/A1/3.
142 Hassell Smith provided evidence of disputes at Stiffkey and Langham from the Bacon papers, which indicate the operation of joint ventures and partnerships in 1604: NRO HBL VIb (5); IVb (4); NRO RAY 257.
143 TM, RAS/A1/3.
144 Rosenheim, *The Townshends of Raynham*, pp. 80–1; Lord Townshend sold Stanhoe to the Cokes in 1681, when he purchased the Toftrees estate, north east of Raynham, from William Ruding.
145 TM, RAS/A1/3.
146 Allison, 'Sheep Corn Husbandry'; H.C. Darby and J. Saltmarsh, 'The Infield-Outfield System on a Norfolk Manor', *Economic History Review*, 3 (1935) 30–44.
147 Accounts for the 1660s, TM, RAD/C3-4; 1670s, RAD/B2-3.
148 Felton's leases: TM, RAS/A1/5.
149 1670s accounts: TM, RAD/B2-3.
150 TM, RAS/A1/3.
151 TM, RAS/A1/3.
152 Peter Stringer's son. TM, RAS/A1/3.
153 TM, RAS/A1/3.
154 TM, RAS/A1/3.
155 TM, RAS/A1/5.
156 D. Stone, 'Productivity and Sheepfarming in late Medieval England', *Agricultural History Review*, 51 (2003) 1–22.
157 TM, RAD/C1: The Rudhams accounted for £379 12s 10d, Stiffkey Hall £389, Elmer's farm at south Raynham £60 and the Morston foldcourse £70.

158 Rate for wintering cows 6/8 each, same as summer grazing; grazing on turnips, 16/- each; ollands 2/- per acre.
159 The other trustees were Sir Jacob Astley, William Thursby and James Calthorpe, Esq of Cockthorpe.
160 TM, RAS/A2/1-8.
161 TM, RAS/A1/6; RAD/C2; RAS/A2/1-8.
162 TM, RAS/A1/6.
163 Rosenheim, *Townshends of Raynham*, pp. 79–81.
164 Ward's correspondence TM, RAS/A1/6; RAS/A1/5.
165 All holdings were re-let by Charles Townshend between 1697–99.
166 Extracts from letting to halves agreements and accounts for the Rudhams in 1693: Appendix 2.
167 TM, RAS/A1/7; RAS/A1/8.
168 See above, pp. 72, 84.

Chapter 6

1 A. Smith, *An Inquiry into the Nature and Causes of the Wealth of Nations*, (1776) Book 3, Part 2, pp. 470–80; A. Young, *Travels Through France During the Years 1787, 1788 and 1789*, ed. C. Maxwell, (Cambridge: Cambridge University Press, 1929), pp. 16, 287.
2 Smith and Young were both strongly influenced by the French physiocrats, notably A.R.G. Turgot, *Reflections on the Formation and Distribution of Wealth*, first published in 1766, ten years before Smith's, *Wealth of Nations*, and Young by his experience in France on the eve of the Revolution.
3 J.S. Mill, *Principles of Political Economy*, (1st edn 1848). References are to the Longman edn 1909, ed. W.J. Ashley, pp. 242–48, 256–82, 203–17.
4 G.E. Mingay, 'The Agricultural Depression, 1730–1750', *Economic History Review*, 2nd series, 8 (1956) 323–38; J.V. Beckett, 'Regional Variation and the Agricultural Depression, 1730–1750', *Economic History Review*, 2nd series, 35 (1982) 35–51; R.A. Dodgshon, 'Coping with Risk: Subsistence Crises in the Scottish Highlands and Islands, 1600–1800', *Rural History*, 1 (2004) 1–25.
5 M. Turner, 'The land tax, land, and property: old debates and new horizons', in M. Turner and D. Mills (eds) *Land Tax and property: The English Land Tax, 1692–1832*, (Gloucester: Alan Sutton, 1986), pp. 1–35.
6 P. Jenkins, *The Making of a Ruling Class: The Glamorgan Gentry, 1640–1790*, (Cambridge: Cambridge University Press, 1983); J.M. Rosenheim, *The Townshends of Raynham: Nobility in Transition in Restoration and Early Hanoverian England*, (Middletown, Ct.: Wesleyan University Press, 1989); idem, *The Emergence of a Ruling Order: English Landed Society 1650–1750*, (Longman, 1998).
7 J.M. Rosenheim, 'County governance and elite withdrawal in Norfolk, 1660–1720' in A.L. Beier, D. Cannadine and J.M. Rosenheim (eds) *The First Modern Society: Essays in English History in Honour of Lawrence Stone*, (Cambridge: Cambridge University Press, 1989), pp. 106–15.
8 M. Overton, *Agricultural Revolution in England: The Transformation of the Agrarian Economy, 1500–1850*, (Cambridge: Cambridge University Press, 1996), pp. 168–78.

9   A. Offer, 'Farm tenure and land values in England, c.1750–1950', *Economic History Review*, 2nd series, 44 (1991), p. 2.

10 D. Stead, 'Risk and risk management in English Agriculture, 1750–1850', *Economic History Review*, 2nd series, 57 (2004), pp. 334–5.

11 R.C. Allen, *Enclosure and the Yeoman: The Agricultural Development of the South Midlands, 1450–1850*, (Oxford: Clarendon Press, 1992).

12 Norfolk Record Office (hereafter NRO), NRS 16338 32C2; E.M. Griffiths, 'The management of two East Norfolk Estates in the Seventeenth Century: Blickling and Felbrigg', (Unpublished PhD thesis, University of East Anglia, 1987), pp. 410, 486.

13 NRO, NRS 16301 32B5; NRO, Trafford, 234 88 X 1. See above, p. 86.

14 Townshend MSS (hereafter TM) RAS/B3/9.

15 BL Add. MSS. 41656/182-183.

16 TM, RAS/B3/4; RAS/A1/7; RAS/A1/8.

17 Unlike his cousin, Ashe Windham, Charles Townshend retained a personal interest in the day-to-day management of his estate, so we have the benefit of much correspondence. He wrote regularly to Ward on plans to improve the park and farms in the Raynham area, planting schemes for the orchard, the disposal of surplus dairy produce and how to deal with difficult tenants – he also revealed his famous enthusiasm for turnips, designating areas to be sown, so much so, that Ward warned of their limitations. He pointed out their unsuitability in certain situations, for example, in the winter when they became frosted and 'so hard the cattel cannot eat them'; in these conditions an ample supply of hay was preferable. TM RAS/B3/4; BL Add. MSS 41655.184.

18 NRO, LEST, KA11. Account of Sir Christopher Calthorpe, trustee for Sir Nicholas Le Strange, 1669–1682. No further evidence of farming to halves was found at Hunstanton, but much material has been lost.

19 TM, RAS/A1/6, RAS/A1/7.

20 TM, RAD/C4.

21 TM, RAS/A3/1-8; RAD/C3-4.

22 R. Santos, 'Risk-sharing and social differentiation of demand in land-tenancy markets in southern Portugal, seventeenth-nineteenth centuries', *Continuity and Change*, 21 (2006) 287–312.

23 Santos, 'Risk-sharing', p. 294.

24 G.E. Mingay, 'The Size of Farms in the Eighteenth Century', *Economic History Review*, 2nd series, 14 (1962) 469–88, estimated the figure at four times the rent. A. Young, *A Farmer's Guide to Hiring and Stocking Farms*, (1770) estimated £300 for a Cheshire Dairy Farm with a herd of 20 cows.

25 R. Mitchison, 'The Old Board of Agriculture, 1793–1822', *English Historical Review*, 74 (1959) 41–69.

26 J. Claridge, *General View of the Agriculture in the County of Dorset*, (1793), pp. 14–15.

27 W. Stevenson, *General View of the Agriculture of the County of Dorset*, (1815), pp. 381–90.

28 Claridge, *General View of the Agriculture in the County of Dorset*, estimated that in 'the poorest parts [rents] were as low as fifty shillings or three pounds a head, and in others as high as six pounds ten shillings, or seven pounds',

p. 15; in one parish rates reached eight pounds. By 1815 the rent per cow had doubled to an average of £12 a cow.

29   *Ibid.* 'The dairies were managed by making all the cream into butter, and from the skimmed milk an inferior kind of cheese, which sells from 25s to 30s per hundredweight ... the butter, worth eight or ten pence per pound, is in general salted down in tubs and supplies Portsmouth and the London markets ... there is also a considerable quantity of better cheese, which brings a price as high as 37s or 2 guineas per hundredweight'.

30   J.F. James and J.H. Bettey (eds) *Farming in Dorset: Diary of James Warne, 1758; Letters of George Boswell, 1787–1805*, Dorset Record Society, Vol. 13 (1993), pp. 12, 29; G.E. Mingay, 'The Diary of James Warne, 1758', *Agricultural History Review*, 38 (1990) 72–8.

31   P. Horn, 'The Dorset dairy system', *Agricultural History Review*, 26 (1978) 100–7.

32   *Ibid.* p. 101.

33   C.A. Horn, 'Two Centuries of Incentive Payments in Agriculture', *The Accountants' Review*, 26 (1975), p. 223.

34   Horn, 'The Dorset Dairy System', p. 106.

35   B. Kerr, *Bound to the Soil*, (John Baker, 1968), pp. 57–8.

36   Moving farms is very much the pattern in New Zealand, where sharemilkers progress from farm to farm, building up their herd size. Eventually, they will sell part of the herd to raise the capital to buy a small farm; then the cycle starts again, progressing to larger farms. In recent years, this cycle has been broken by the rising price of agricultural land; many opt to remain sharemilkers investing surplus capital in very large herds or urban property, which yield better returns.

37   W. Marshall, *The Rural Economy of the West of England*, Vol. 2, (1796), pp. 150–1.

38   G.V. Harrison, 'The South West: Dorset, Somerset, Devon, and Cornwall', in J. Thirsk (ed.) *The Agrarian History of England Wales, Vol. V, 1640–1750. I. Regional Farming Systems* (Cambridge: Cambridge University Press, 1984), p. 378 pointed out that Horn was wrong in assuming that the practice was confined to Dorset.

39   J. Broad, 'Regional Perspectives and variations in English dairying, 1650–1850', in R.W. Hoyle (ed.) *People, Landscape and Alternative Agriculture: Essays for Joan Thirsk*, Agricultural History Review, Supplement 3 (2004), p. 103, refers to the disappearance of dairying in East Anglia, most particularly, Suffolk Cheese by 1800.

40   A.R. Wilson, *Forgotten Harvest: The Story of Cheesemaking in Wiltshire*, (Calne: the author, 1995), refers to the practice as sharemilking, which reflects her experience in the United States dairy industry and her familiarity with such systems.

41   T. Davies, *General View of the Agriculture of Wiltshire*, (1811), pp. 135–6.

42   G.B. Worgan, *General View of the Agriculture of the County of Cornwall*, (1811), p. 141.

43   J. Carmona, 'Sharecropping and livestock specialization in France, 1830–1930', *Continuity and Change*, 21 (2006) 235–59.

44   Santos, 'Risk-sharing', pp. 287–312.

45   S. Ogilvie, '"Whatever is, is right?" Economic institutions in pre-industrial Europe', *Economic History Review*, 2nd series, 60 (2007) 649–84.

46 Stead, 'Risk and risk management', p. 334; Offer, 'Farm tenure and land values', p. 2.
47 Mingay, 'Agricultural Depression', pp. 323–38.
48 Stead, 'Risk and risk management ', pp. 334–61.
49 Santos, 'Risk-sharing', pp. 287–312.

Chapter 7

1 The most well-known publication in this vein is W. Cobbett, *Rural Rides*, (1830).
2 There is an extensive literature on the fate of the rural labourer in the late eighteenth and early nineteenth centuries: see M. Overton, *Agricultural Revolution in England: The Transformation of the Agrarian Economy 1500–1850*, (Cambridge: Cambridge University Press, 1996), pp. 182–91. More recent contributions include, L. Shaw-Taylor, 'Labourers, Cows, Common Rights and Parliamentary Enclosure: The Evidence of Contemporary Comment c. 1760–1810', *Past & Present*, 171 (2001) 95–126; and *idem*, 'Parliamentary Enclosure and the Emergence of an English Agricultural Proletariat', *Journal of Economic History*, 61 (2001) 640–62.
3 J.S. Mill, *Principles of Political Economy*, (1st edn 1848). References are to the Longman edn 1909, ed. W.J. Ashley, p. 316.
4 D. Stead, 'Risk and risk management in English Agriculture, 1750–1850', *Economic History Review*, 2nd series, 57 (2004) 334–61.
5 M. Reed, 'The peasantry of nineteenth century England: a neglected class?', *History Workshop Journal*, 18 (1984) 53–76.
6 A. Howkins, 'Social, cultural and domestic life', in E.J.T. Collins (ed.) *The Agrarian History of England and Wales, Vol. VII, 1850–1914, part II*, (Cambridge: Cambridge University Press, 2000), p. 1508.
7 M. Overton, 'Agriculture', in J. Langton and R.J. Morris (eds) *Atlas of Industrializing Britain*, (Methuen, 1986), p. 46; 'Agricultural Returns for Great Britain for 1895, BPP 1896, XCII, c.8073; G. Shaw-Lefevre, *Agrarian Tenures: A Survey of the Laws and Customs Relating to the Holding of Land in England, Ireland and Scotland*, (Cassell, 1893). In his Return of Land-owners, 1870, out of a total of 33m acres in the hands of lay landowners 2,250 landowners owned 50 per cent of the total with estates over 2,000 acres, averaging 7,300 acres each; 1,750 landowners owned 2.5m. acres with estates between 1,000 and 2,000 acres; 34,000 landowners owned c.9m acres with estates between 100 and 1,000 acres; 217,000 landowners owned c.4m acres with estates between 1 and 100 acres.
8 Reed, 'Peasantry of nineteenth century England', p. 54, citing, L. de Lavergne, *The Rural Economy of England, Scotland and Ireland*, (Paris: Guillaumin, 1858), p. 108.
9 A. Young, *General Report on Enclosures*, (1808), pp. 32–4.
10 A. Young, *General View of the Agriculture of the County of Norfolk*, (1804), p. 107.
11 *Ibid.* pp. 138–9.
12 *Ibid.* p. 134.
13 *Ibid.* p. 125.

14   A.J. Peacock, *Bread or Blood: A Study of the Agrarian Riots in East Anglia in 1816*, (Gollancz, 1965).

15   E.J. Hobsbawm and G. Rudé, *Captain Swing*, (Lawrence & Wishart, 1969).

16   J. Burchardt, *The Allotment Movement in England, 1793–1873*, (Woodbridge: Boydell, 2002).

17   Mill, *Principles of Political Economy*, pp. 368–9.

18   See above, pp. 121–5.

19   S. Taylor, *Profit-sharing between Capital and Labour*, (Kegan Paul, 1884).

20   *Ibid.* pp. 102–8; M. Chase, '*The People's Farm': English Radical Agrarianism, 1775–1840*, (Oxford: Clarendon, 1988), pp. 153–4. Craig formerly editor of the *Lancashire and Yorkshire Co-operator*, ensured that the experiment received maximum publicity in the co-operative press. See also R.G. Garnett, *Co-operation and the Owenite Socialist Communities in Britain, 1825–45*, (Manchester: Manchester University Press, 1972), pp. 41–64 for a full description of the Ralahine experiment; C. Ó Gráda, 'The Owenite Community at Ralahine, County Clare, 1831–1833: A Reassessment', *Irish Economic and Social History*, 1 (1974) 36–48.

21   Taylor, *Profit-sharing*, p. 103.

22   J. MacAskill, 'The Chartist Land Plan' in A. Briggs (ed.) *Chartist Studies*, (Macmillan, 1959), pp. 304–41 and A.M. Hadfield, *The Chartist Land Company*, (Newton Abbot: David & Charles, 1970) remain the only extended discussions. M. Chase, '"We only wish to work for ourselves": the Chartist Land Plan', in M. Chase and I. Dyck (eds) *Living and Learning: Essays in Honour of J.F.C. Harrison*, (Aldershot: Scolar, 1996), pp. 133–48 brings these discussions up to date.

23   A. Howkins, 'Types of rural communities', in E.J.T. Collins (ed.) *The Agrarian History of England and Wales, Vol. VII, 1850–1914, part II*, (Cambridge: Cambridge University Press, 2000), pp. 1334–7, reminds us the Chartist settlement at Great Dodford, Bromsgrove survived until the first World War, selling fruit and vegetables.

24   Mill, *Principles of Political Economy*, pp. 302–17.

25   *Ibid.* pp. 315, 307.

26   *Ibid.* pp. 307–8.

27   Mill, *Principles of Political Economy*, p. 308. F.M. Snowden, *The Fascist Revolution in Tuscany, 1919–1922*, (Cambridge: Cambridge University Press, 1989), pp. 7–69, offers a broader perspective on the virtues of *mezzadria*.

28   Mill, *Principles of Political Economy*, p. 316.

29   *Ibid.* pp. 265–6.

30   See below, pp. 121–5.

31   J.R. McQuiston, 'Tenant Right: Farmer against Landlord in Victorian England, 1847–1883', *Agricultural History*, 47 (1973) 95–113.

32   W.O. Massingberd, *History of the Parish of Ormsby-cum-Ketsby*, (Lincoln, 1893), pp. 303–4.

33   J.A. Perkins, 'Tenure, Tenant Right and Agricultural Progress in Lindsey, 1780–1850', *Agricultural History Review*, 23 (1975) 1–22.

34   See above, pp. 52–3.

35   E.J.T. Collins, 'Food supplies and food policy', in *idem* (ed.) *The Agrarian History of England and Wales, Vol. VII, 1850–1914, part I*, (Cambridge: Cambridge University Press, 2000), pp. 33–71.

36 J.R. Fisher, 'Agrarian politics', in E.J.T. Collins (ed.) *The Agrarian History of England and Wales, Vol. VII, 1850–1914, part I*, (Cambridge: Cambridge University Press, 2000), pp. 346–8.

37 J. Marsh, *Back to the Land: the Pastoral Impulse in England from 1880–1914*, (Quartet Books, 1982); P. Gould, *Early Green Politics, 1880–1900*, (Brighton: Harvester Press, 1988); Howkins, 'Types of rural communities', pp. 1334–7.

38 J. Habakkuk, *Marriage, Debt, and the Estates System: English Landownership 1650–1950*, (Oxford: Clarendon Press, 1994), pp. 645–9.

39 Shaw-Lefevre, *Agrarian Tenures*, pp. 1–31; G.C. Broderick, *English Land and English Landlords*, (Cassell, 1881), pp. 434–5.

40 J.E. Thorold Rogers, *Six Centuries of Work and Wages: the History of English Labour*, (11th edn, T. Fisher Unwin, 1912), p. 279.

41 Shaw-Lefevre, *Agrarian Tenures*. G.J. Shaw-Lefevre, 1st Baron Eversley, (1831–1928), was Liberal M.P. for Reading 1863–1885 and Bradford 1885–1895. He was a land reformer, strongly influenced by J.S. Mill. A founder member of the Commons Preservation Society (1865), he secured the preservation of royal forests, metropolitan and rural commons for public benefit. He was Chairman of the Select Committee on Irish Land Act 1875; Chairman of the Royal Commission into the depression in agriculture, 1894–95. A. Warren, 'Lefevre, George John Shaw, Baron Eversley (1831–1928)', *Oxford Dictionary of National Biography*, (Oxford: Oxford University Press, 2004), online edn, Jan 2008 [http://0-www.oxforddnb.com.lib.ex.ac.uk:80/view/article/36055, accessed 9 Aug 2008].

42 A. Offer, 'Farm tenure and land values in England, c.1750–1950', *Economic History Review*, 2nd series, 44 (1991) 1–20.

43 Shaw-Lefevre, *Agrarian Tenures*, pp. 1–34.

44 E.H. Hunt and S.J. Pam, 'Responding to agricultural depression, 1873–96, managerial success, entrepreneurial failure', *Agricultural History Review*, 50 (2003) 225–52, reiterate the problems associated with a rigid and polarized landed society, which made co-operation difficult and diversification into "small men's crops", such as poultry, almost unthinkable.

45 R.J. Loyd-Lindsay, 1st Baron Wantage, (1832–1901), soldier and agriculturist, married Sarah Loyd, daughter of 1st Baron Overstone, who settled on the couple the Lockinge Estate, Wantage, Berkshire. By 1874, it was the largest estate in Berkshire, and by 1883 comprised 52,000 acres. He was a Tory M.P. 1865–1885; created Baron Wantage in 1885, and in 1886 became Lord Lieutenant of Berkshire. He was related by marriage to G. Shaw Lefevre; R.T. Stearn, 'Lindsay, Robert James Loyd, Baron Wantage (1832–1901)', *Oxford Dictionary of National Biography*, (Oxford: Oxford University Press, 2004), online edn, May 2008 [http://www.oxforddnb.com/view/article/ 34544, accessed 9 Aug 2008]. See M.A. Havinden, *Estate Villages: a study of the Berkshire villages of Ardington and Lockinge*, (Reading: Lund Humphries, 1966) and also, Howkins, 'Types of rural communities', p. 1321, for a more sceptical view of created communities.

46 Royal Commission on Labour Reports by C.M. Chapman, Berkshire, BPP, 1893–4, XXXV, c.6894-II, pp. 61–73.

47 Havinden, *Estate Villages*; A. Grey, 'Profit Sharing in Agriculture', *Journal of the Royal Agricultural Society of England*, 3rd series, 2 (1891) pp. 779–80; D.R. Mills, *Lord and Peasant in Nineteenth-Century Britain*, (Croom Helm, 1980), pp. 31–8.

48 Albert Grey, 4th Earl Grey, (1851–1917), Liberal M.P. for S. Northumberland, 1880–85 and Tyneside 1885–89; Lord Lieutenant of Northumberland 1889–1904; Governor-General of Canada 1904–11. He is described as a reformer and idealistic imperialist. In the commons he was leader of the Grey Committee, which advocated proportional representation, consumer and industrial co-operatives, industrial profit-sharing, church reform, temperance and the garden city movement: i.e. 'the reconstruction of national life'. He succeeded to the title in 1894, after having managed his uncle's estate, of about 17,600 acres, since 1884. Related by marriage to Lady Wantage and G.J. Shaw-Lefevre; C. Miller, 'Grey, Albert Henry George, fourth Earl Grey (1851–1917)', *Oxford Dictionary of National Biography*, Oxford University Press, Sept 2004; online edn, Jan 2008 [http://www. oxforddnb.com/view/article/33568, accessed 9 Aug 2008].

49 Grey, 'Profit Sharing in Agriculture'; Royal Commission on Labour, Reports by A. Wilson Fox, Northumberland, BPP, 1893–4, XXXV, c.6894-III, pp. 108–10, 116, 132–3; see above, Table 2.1, p. 14.

50 In Cambridgeshire, Lord George Manners of Cheveley Park, 'experimented with partnership farming between 1872–1874, dividing part of the profits of Ditton Lodge Farm among his labourers'. C.P. Lewis, 'Woodditton', in A.F. Wareham and A.P.M. Wright (eds) *A History of the County of Cambridge and the Isle of Ely, Vol. X, North-eastern Cambridgeshire*, (Oxford: Oxford University Press for the Institute of Historical Research, 2002), p. 92. From the *Cambridge Chronicle*, 10 Jan 1874, p. 8; 19 Sept. 1874, p. 7; 12 Dec. 1874, p. 4.

51 Grey, 'Profit sharing in agriculture', p. 772, n. 1.

52 *Ibid.* pp. 781–91.

53 *Ibid.* p. 781.

54 *Ibid.* pp. 780–1, n. 1.

55 M. Betham-Edwards (ed.) *Arthur Young: Travels Through France*, 4th edn (1892); *idem, France of Today, A Survey Comparative and Retrospective*, (1892), pp. 210–11.

56 *Ibid.* pp. vi–viii.

57 H. Higgs, 'Métayage in Western France', *The Economic Journal*, 4 (1894) 1–13.

58 J. Carmona, 'Sharecropping and livestock specialization in France, 1830–1930', *Continuity and Change*, 21 (2006) 235–59.

59 See above, p. 90.

60 R.E. Prothero, *The Pleasant Land of France*, (1908), pp. 104–14.

61 J. Thirsk, *Alternative Agriculture: A History from the Black Death to the Present Day* (Oxford: Clarendon Press, 1997), pp. 204–16.

62 M. Freeman, *Social Investigation and Rural England*, 1870–1914, (Woodbridge: Boydell Press, 2002), pp. 168–82.

63 Mills, *Lord and Peasant in Nineteenth-Century Britain*.

64 K. Sinclair, *A History of New Zealand* (Penguin, 2000), pp. 93–114.

65 Similarly, in Australia, from the 1840s, reflecting the labour shortage and increase in flock size, shepherds received incentive payments often amounting to a third of the lambs born. This arrangement is reminiscent of the practice at Raynham, Norfolk in the 1620s and 1630s. J. Pickard, 'Shepherding in Colonial Australia', *Rural History*, 19 (2008), p. 63.

66 Sinclair, *A History of New Zealand*, p. 99.
67 H. George, *Progress and Poverty*, (Reeves, 1880); A.R. Wallace, *Land Nationalization: its Necessity and Aims*, (Trubner, 1882); Sinclair, *A History of New Zealand*, pp. 173–8.
68 Sinclair, *A History of New Zealand*, pp. 185–90; T. Booking, 'Economic transformation', in G.W. Rice (ed.) *The Oxford History of New Zealand*, (Oxford: Clarendon Press, 1992), pp. 233–43.
69 R.W.M. Johnson, *Reforming EU Farm Policy: Lessons from New Zealand*, (Institute of Economic Affairs, 2000); between 1984 and 1997 subsidies as a proportion of farm output were reduced from 32.8 per cent to 2.3 per cent.
70 F.J. Hall and P. Martyn, *Changes in Sharemilking: 1973–1993*, Technical Paper, 11 (Ministry of Agriculture, New Zealand, 1993); G. Taylor, *A Review of Sharemilking: 1972–1996*, (Rural Policy Unit, Ministry of Agriculture, New Zealand, 1996).
71 C. Blakeney, *Joint Ventures in Dairy Farming*, (Nuffield Farm Scholarship Trust, 1995), pp. 1–11.
72 *Ibid.* p. 18.

Chapter 8

 1 Table 2.1, above, p. 14.
 2 J.A. Venn, *Foundations of Agricultural Economics*, (Cambridge: Cambridge University Press, 1923), pp. 43–7.
 3 Hereafter CLA. It was founded in 1907 and became the Country Land and Business Association in 2001 but maintains the title CLA.
 4 See above, p. 6.
 5 G.E. Fussell (ed.) *Robert Loder's Farm Accounts 1610–1620*, Camden Society, 3rd series, Vol. 53 (1936).
 6 W.C. Cooke (1858–1931), Coppice Farm, Ratlinghope, Shropshire.
 7 Ponies were reared for the coal pits and sold at Church Stretton Horse Fair, R. Perren, 'Agriculture 1875–1985, in G.C. Baugh (ed.) *The Victoria History of the County of Shropshire, Vol. 4, Agriculture*, (Oxford: Oxford University Press for the Institute of Historical Research, 1989), pp. 232–69.
 8 B. Price of Bockleton Court, Stoke St. Milborough, Shropshire.
 9 B. Davies, Penywern, Clun, Shropshire.
10 *Wellington Journal* and *Shrewsbury News*, 28 November 1931.
11 J. Bevan, sister of Edward Foster, Newton House, Bridgnorth, Shropshire.
12 See above, pp. 16, 47–51, and Jim Lewis, Liskeard, Cornwall.
13 *Hereford Times*, Sept 4th, 1959.
14 G. Henderson, *The Farming Ladder*, (Faber & Faber, 1943).
15 R. Stanes, *The Old Farm: A History of Farming Life in the West Country*, (Exeter: Devon Books, 1990), p. 7.
16 Correspondence with R. Stanes.
17 J. Keith, *Fifty Years of Farming*, (Faber & Faber, 1954), p. 42.
18 R. Walton and M. Essayon, *Adkins's Landlord and Tenant*, 18th edn (The Estates Gazette Ltd, 1982), pp. 133–4; H.A.C. Densham, *Scammell and Densham's Law of Agricultural Holdings*, 7th edn (Butterworth, 1989), pp. 120–1.

19   R. Perren, *Agriculture in Depression 1870–1940*, (Cambridge: Cambridge University Press, 1995), pp. 68–70.

20   E.J.T. Collins, 'Rural and agricultural change', in *idem* (ed.) *The Agrarian History of England and Wales, Vol. VII, 1850–1914, part I*, (Cambridge: Cambridge University Press, 2000), pp. 208–23; A.D. Hall, *A Pilgrimage of British Farming, 1910–1912*, (Murray, 1913), pp. 205–18.

21   Letter to Benjamin Griffiths from Doolittle & Dalley, Land Agents, Kidderminster, 23 March 1921, private collection.

22   Perren, *Agriculture in Depression*, p. 36; S.G. Sturmey, 'Owner-farming in England and Wales 1900–1950', *Manchester School*, 23 (1955) 246–68.

23   The Agricultural Holdings Act 1883 guaranteed the right of tenants to unexhausted improvements; these rights were consolidated under the Acts of 1980 and 1923. The Agricultural Holdings Act 1948 granted tenants security of tenure for their lifetime, which was extended in 1976 for a further two generations.

24   Sturmey, 'Owner-Farming', pp. 246–68.

25   In the agricultural census of 1988, 37 per cent of agricultural land in England and Wales was tenanted, the Royal Institute of Chartered Surveyors calculated the figure at 41 per cent.

26   Sturmey, 'Owner Farming', p. 256.

27   W. Marshall, *The Rural Economy of Norfolk*, Vol. 1, (1787), p. 2.

28   A.J. Papworth, 'Diversity and Diversification', *Journal of the Royal Agricultural Society of England*, 164 (2003) [http://www.rase.org.uk/activities/publications/RASE_journal/2003/09-58151622.pdf].

29   Jim Papworth was born in 1941, his brother David in 1945.

30   Now part of Anglia Farmers, an amalgamation of similar groups formed in East Anglia at this time.

31   P. Barnes, *Norfolk Landowners since 1880*, (Norwich: Centre of East Anglian Studies, 1993).

32   Their need for grass has extended to the point where they hire 120 hectares, reminding us of livestock farmers in the Welsh Borders hiring grazing for sheep and cattle at tack, and placing ewes at halves.

33   The Agricultural Holdings Act 1883 guaranteed tenants rights for unexhausted improvements. Agricultural Holdings Act of 1948 provided tenants with full life time security of tenure, which was extended to three generations in 1976. R. Gibberd, 'Farm Tenancies: Paradise Leased or Paradise Lost', *Journal of the Royal Agricultural Society of England*, 156 (1995) 106–13. M. Winter, 'Revisiting landownership and property rights', in H. Clout (ed.) *Contemporary Rural Geographies, Land, Property and Resources in Britain: Essay in Honour of Richard Munton*, (University College London Press, 2007), pp. 72–83.

34   R. Stratton, *Sharefarming*, (Country Landowners Association, 1983).

35   The Act abandoned security of tenure and instigated leases for terms of years agreed by the two parties, with 12 months notice to quit on either side. The agreement should be individual to the parties, and negotiated by them to suit their particular conditions, rather than hedged about with restrictive social legislation. To encourage landowners further, they gained concessions over tax, notably 100 per cent relief on Inheritance Tax. At the same time, the Act guaranteed tenants protection for their improvements.

36 M. Winter, C. Richardson, C. Short and C. Watkins, *Agricultural Land Tenure in England and Wales*, (Royal Institution of Chartered Surveyors, 1990).

37 See below, pp. 180–93.

38 Letter from Peter Fletcher to Liz Griffiths, 16 July 1998.

39 Despite John Young's reservations, the National Trust entered into a new Share Farm Agreement with the tenants at Hardwick Hall in 2005.

40 Communication from W. Gemmill of Strutt and Parker. In 2008 Strutt and Parker's website includes ten experts on Contract Farming and Joint Ventures; [http://wai.struttandparker.co.uk/html/contract-farming.php, accessed 11 August 2008].

41 J. Hunt, 'Landowner Profile', *CLA Magazine*, (Country Landowners Association, 2005) p. 4.

42 Pers. comm. July 1998.

43 Ian MacNicol was interviewed at Stody Lodge in 2002, he died in 2006.

44 M. Little, 'Joint and several: Dealing with farm agreements post CAP reform', *Savills Aspects of Land East*, (Autumn/Winter 2005).

45 The Single Payment Scheme is the principal agricultural subsidy scheme in the European Union.

46 W. Gemmill, 'The role of the new entrant', *Journal of the Royal Agricultural Society of England*, 165 (2005).

47 'Herd the news at Hardwick?', *East Midland News*, (The National Trust, Autumn & Winter 2005/6).

Chapter 9

1 *A Century of Agricultural Statistics: Great Britain 1866–1966*, (HMSO, 1968).

2 J. Murdoch and N. Ward, 'Governmentality and territoriality: the statistical manufacture of Britain's "national farm"', *Political Geography*, 16 (1997) 307–24; B. Short, C. Watkins, W. Foot and P. Kinsman, *The National Farm Survey 1941–1943: State Surveillance and the Countryside in England and Wales in the Second World War*, (Wallingford: CABI, 2000).

3 M. Winter, C. Richardson, C. Short and C. Watkins, *Agricultural Land Tenure in England and Wales*, (Royal Institution of Chartered Surveyors, 1990).

4 B. Hill and R. Gasson, 'Farm tenure and farming practice', *Journal of Agricultural Economics*, 36 (1985) 187–99; D. Cannadine, *The Decline and Fall of the British Aristocracy*, (New York: Yale University Press, 1990); D.I. Bateman, 'Heroes for present purposes? – a look at the changing idea of communal land ownership in Britain', *Journal of Agricultural Economics*, 40 (1989) 269–89.

5 Short, Watkins, Foot, Kinsman, *The National Farm Survey 1941–1943*.

6 D. Rose, H. Newby, P. Saunders and C. Bell, 'Land tenure and official statistics: a research note', *Journal of Agricultural Economics*, 28 (1977) 69–75; H. Newby, C. Bell, D. Rose and P. Saunders, *Property Paternalism and Power*, (Hutchinson, 1978).

7 Winter, Richardson, Short, Watkins, *Agricultural Land Tenure*.

8 H.A.C. Densham, *Scammell and Densham's Law of Agricultural Holdings*, (Butterworth, 1989).

9 J. Nix, P. Hill and N. Williams, *Land and Estate Management*, (Chichester: Packard, 1987); T. Marsden, 'Property-state relations in the 1980s: an

examination of landlord-tenant legislation in British agriculture', in G. Cox, P. Lowe, and M. Winter (eds) *Agriculture: People and Policies*, (Allen and Unwin, 1986), pp. 126–45; D.A.G. Troup, *Agricultural Holdings Act, 1984: The Practitioner's Companion*, (Surveyor's Publication, 1984).

10   R. Stratton, *Sharefarming*, (Country Landowners Association, 1983).
11   M. Winter, 'Revisiting landownership and property rights', in H. Clout (ed.) *Contemporary Rural Geographies, Land, Property and Resources in Britain: Essay in Honour of Richard Munton*, (University College London Press, 2007), pp. 72–83.
12   Winter, Richardson Short, and Watkins, *Agricultural Land Tenure*.
13   I. Whitehead, A. Errington, N. Millard and T. Felton, *An Economic Evaluation of the Agricultural Tenancies Act 1995*, (University of Plymouth Report to DEFRA, 2002).
14   R. Gibbard, N. Ravenscroft and J. Reeves, 'The Popular Culture of Agricultural Law Reform', *Journal of Rural Studies* 15 (1999) 269–78.
15   *Study of Joint Venture Farming. Report to DEFRA*, (Wolverhampton: ADAS, 2007); A. Butler and M. Winter, *Agricultural Land Tenure in England and Wales 2008*, University of Exeter, Centre for Rural Policy Research Paper No. 24 (2008).
16   *Study of Joint Venture Farming*, p. 7.
17   *Ibid.*
18   Type A in our taxonomy shown in Table 2.1, above, p. 14.
19   *Study of Joint Venture Farming*, p. 9.
20   The areas are for England and Wales, extrapolated from the sample.
21   *Tenanted Land Survey – England 2006*, DEFRA stats 09/07 [https://statistics. defra.gov.uk/esg/statnot/astl.pdf] accessed 11 August 2008.
22   There is a statistical association between increasing land in the past five years because of farm profitability and having a sharefarming agreement.
23   See above, pp. 12, 15–16, 153–4.
24   See above, pp. 49–51.

Chapter 10

1   See Table 2.1, above, p. 14.
2   *The Archers*, broadcast on BBC Radio 4, 20 March 2008.
3   J. Carmona and J. Simpson, 'Why sharecropping?: explaining its presence and absence in Europe's vineyards, 1750–1950' *Universidad Carlos III de Madrid, Working Papers in Economic History*, 11 (2007), argues that sharecropping in France was much less common than traditionally assumed.
4   See above, pp. 126–8.

Bibliography

The place of publication is London unless otherwise stated

A Century of Agricultural Statistics: Great Britain 1866–1966 (1968) (HMSO).

Agricultural Returns for Great Britain for 1895, BPP 1896, XCII, c.8073.

Allen, R.C. (1992) *Enclosure and the Yeoman: The Agricultural Development of the South Midlands, 1450–1850*, (Oxford: Clarendon Press).

Allison, K.J. (1957) 'The Sheep-corn Husbandry of Norfolk in the Sixteenth and Seventeenth Centuries', *Agricultural History Review*, 5, 12–30.

Anon. 'Herd the news at Hardwick?', *East Midland News*, (The National Trust, Autumn & Winter 2005/6).

Antoine, A. (2003) 'La Légende noire du métayage dans l'ouest de la France (XVIII–XIX siècle)', in Béaur, G., Arnoux, M. and Varet-Vitu, A. (eds) 'Exploiter la Terre: Les Contrats agraires de l'Antique à nos jours, Actes du colloque international, Caen, 1997', *Bibliothèque d'Histoire Rurale*, 7, 457–70.

Antoine, A. (forthcoming) 'Métayage, productivité et économie monétaire. Les enseignements de l'analyse des comptabilités agricoles', *Agricultural History Review*.

Barnes, P. (1993) *Norfolk Landowners since 1880*, (Norwich: Centre of East Anglian Studies).

Bateman, D.I. (1989) 'Heroes for Present Purposes? – A Look at the changing Idea of Communal Land Ownership in Britain', *Journal of Agricultural Economics*, 40, 269–89.

Beckett, J.V. (1982) 'Regional Variation and the Agricultural Depression, 1730–1750', *Economic History Review*, 2nd series, 35, 35–51.

Betham-Edwards, M. (1892) *France of Today, A Survey Comparative and Retrospective*.

Betham-Edwards, M. (ed.) (1892) *Arthur Young: Travels Through France*.

Bettey, J.H. (2003) 'The Development of Water Meadows on the Salisbury Avon, 1665–1690', *Agricultural History Review*, 51, 163–72.

Bettey, J.H. (2005) *Wiltshire Farming in the Seventeenth Century*, Wiltshire Record Society, vol. 57.

Bettey, J.H. (2007) 'The Floated Water-Meadows of Wessex: A Triumph of English Agriculture', in Cook, H. and Williamson, T. (eds) *Water Meadows: History, Ecology and Conservation*, (Macclesfield: Windgather Press).

Blakeney, C. (1995) *Joint Ventures in Dairy Farming*, (Nuffield Farm Scholarship Trust).

Bloch, M. (1966) *Rural French Society*, (Routledge and Kegan Paul).

Blomefield, F. (1805–10) *An Essay Towards a Topographical History of the County of Norfolk*, 11 vols, first published 1739–75.

Blum, J. (1978) *The End of the Old Order in Rural Europe*, (Princeton: Princeton University Press).

Blunder, G., Moran, W., and Bradly, A. (1997) 'Archaic relations of Production in Modern Agricultural Systems: The Example of Sharemilking in New Zealand', *Environment and Planning A*, 29, 1759–76.

Booking, T. (1992) 'Economic Transformation', in Rice, G.W. (ed.) *The Oxford History of New Zealand*, (Oxford: Clarendon Press), pp. 233–43.

Britnell, R.H. (1988) 'The Pastons and Their Norfolk', *Agricultural History Review*, 36, 132–44.

Britnell, R.H. (1991) 'Farming Practice and Techniques: Eastern England', in Miller (ed.) *Agrarian History of England and Wales, Vol. III, 1348–1500*, Cambridge: Cambridge University Press), pp. 194–209.

Britnell, R.H. (1991) 'The Occupation of the Land: Eastern England', in Miller, E. (ed.) *The Agrarian History of England and Wales vol. III, 1348–1500*, (Cambridge: Cambridge University Press), pp. 53–66.

Britnell, R.H. (2004) *Britain and Ireland 1050–1530: Economy and Society*, (Oxford: Oxford University Press).

Broad, J. (2004) 'Regional Perspectives and Variations in English Dairying, 1650–1850', in Hoyle, R.W. (ed.) *People, Landscape and Alternative Agriculture: Essays for Joan Thirsk*, Agricultural History Review, Supplement 3 (2004), pp. 93–112.

Broad, J. (2004) *Transforming English Rural Society: The Verneys and the Claydons, 1600–1820*, (Cambridge: Cambridge University Press).

Broderick, G.C. (1881) *English Land and English Landlords*, (Cassell).

Browne, T. (1658) *The Garden of Cyrus*.

Burchardt, J. (2002) *The Allotment Movement in England, 1793–1873*, (Woodbridge: Boydell Press).

Butler, A. and Winter, M. (2008) *Agricultural Land Tenure in England and Wales, 2008*, University of Exeter, Centre for Rural Policy Research Paper No. 24.

Câmara, B. (2006) 'The Portuguese Civil Code and the colonia tenancy contract in Madeira (1867–1967)', *Continuity and Change*, 21, 213–33.

Campbell, B.M.S. (1983) 'Agricultural Progress in Medieval England: Some Evidence from Eastern Norfolk' *Economic History Review*, 2nd series, 36, 26–46.

Campbell, B.M.S. (2005) 'The Agrarian Problem in the Early Fourteenth Century', *Past & Present*, 188, 3–70.

Campbell, B.M.S. and Overton, M. (1993) 'A New Perspective on Medieval and Early Modern Agriculture: Six Centuries of Norfolk Farming, c.1250–c.1850', *Past & Present*, 141, 38–105.

Campbell, L. (1990) *Sir Roger Townshend and his Family: A Study of Gentry Life in Early Seventeenth Century Norfolk*, (Norwich: Centre of East Anglian Studies).

Campling, A. (ed.) (1943) *East Anglian Pedigrees*, Norfolk Record Society, vol. 13.

Cannadine, D. (1990) *The Decline and Fall of the British Aristocracy*, (New York: Yale University Press).

Carmona, J. (2006) 'Sharecropping and Livestock Specialization in France, 1830–1930', *Continuity and Change*, 21, 235–59.

Carmona, J. and Simpson, J. (1999) 'The "Rabassa Morta" in Catalan Viticulture: The Rise and Decline of a Long-Term Sharecropping Contract, 1670s–1920s', *Journal of Economic History*, 59, 290–315.

Carmona, J. and Simpson, J. (2007) 'Why Sharecropping?: Explaining its Presence and Absence in Europe's Vineyards, 1750–1950' *Universidad Carlos III de Madrid, Working Papers in Economic History*, 11.

Carrick, M. (1992) 'Lords of the Manor of Wawne: The Ashe, Windham and Smith Families 1651–1950', *East Yorkshire Local History Society*, 47, 15–21.

Chalklin, C.W. (1965) *Seventeenth-Century Kent*, (Longmans).

Chase, M. (1988) *'The People's Farm': English Radical Agrarianism, 1775–1840*, (Oxford: Clarendon).

Chase, M. (1996) '"We Only Wish to Work for Ourselves": The Chartist Land Plan', in Chase, M. and Dyck, I. (eds) *Living and Learning: Essays in Honour of J.F.C. Harrison*, (Aldershot: Scolar), pp. 133–48.

Chayanov, A.V. (1966) *The Theory of Peasant Farming*, ed. Thorner, D. Kerbaly, B., and Smith, R.E.F. (Homewood, Ill.: American Economic Association).

Chinnery, G.A. (1953) *A Handlist of the Cholmondeley (Houghton) Manuscripts: Sir Robert Walpole's Archive*, (Cambridge: Cambridge University Press).

Claridge, J. (1793) *General View of the Agriculture in the County of Dorset*.

Clark, E. (1982) 'Some Aspects of Social Security in Medieval England', *Journal of Family History*, 7, 307–20.

Clark, G. (2004) 'The Price History of English Agriculture, 1209–1914', *Research in Economic History*, 22, 41–124.

Cobbett, W. (1830) *Rural Rides*.

Cohen, J. and Galassi, F.L. (1990) 'Sharecropping and Productivity: Feudal Residues in Italian Agriculture, 1911', *Economic History Review*, 2nd series, 43, 645–56.

Cohen, J. and Galassi, F.L. (1994) 'The Economics of Tenancy in Early Twentieth Century Southern Italy', *Economic History Review*, 2nd series, 47, 585–600.

Cohen, J.S. (1989) 'Institutions and Economic Analysis', in T.G. Rawksi *et al.* (eds) *Economics and the Historian*, (Berkeley: University of California Press), pp. 60–84.

Collins, E.J.T. (2000) 'Food Supplies and Food Policy', in *idem* (ed.) *The Agrarian History of England and Wales vol. VII, 1850–1914, part I*, (Cambridge: Cambridge University Press), pp. 33–71.

Collins, E.J.T. (2000) 'Rural and Agricultural Change', in *idem* (ed.) *The Agrarian History of England and Wales, Vol. VII, 1850–1914, part I*, (Cambridge: Cambridge University Press).

Darby, H.C. and Saltmarsh, J. (1935) 'The Infield-Outfield System on a Norfolk Manor', *Economic History Review*, 3, 30–44.

Davies, T. (1811) *General View of the Agriculture of Wiltshire*.

De Lavergne, L. (1858) *The Rural Economy of England, Scotland and Ireland*, (Paris: Guillaumin).

De Serres, O. (1600) *Le Théâtre d'Agriculture*, (Paris).

Densham, H.A.C. (1989) *Scammell and Densham's Law of Agricultural Holdings*, 7th edn (Butterworth).

Dodgshon, R.A. (1981) *Land and Society in Early Scotland*, (Oxford: Clarendon Press).

Dodgshon, R.A. (2004) 'Coping with Risk: Subsistence Crises in the Scottish Highlands and Islands, 1600–1800', *Rural History*, 1, 1–25.

Dyer, C.C. (1991) 'Farming Practice and Techniques: The West Midlands', in Miller, E. (ed.) *The Agrarian History of England and Wales, Vol. III, 1348–1500*, (Cambridge: Cambridge University Press), pp. 222–38.

Edwards, P. (1989) 'Agriculture 1540–1750', in Baugh, G.C. (ed.) *The Victoria History of the County of Shropshire, Vol. 4, Agriculture*, (Oxford: Oxford University Press for the Institute of Historical Research), pp. 119–68.

Edwards, P. (1991) *Farming: Sources for Local Historians*, (Batsford).

Ellis, F. (1993) *Peasant Economics: Farm Household and Agrarian Development*, (Cambridge: Cambridge University Press).

Emery, F.V. (1967) 'The Farming Regions of Wales', in Thirsk, J. (ed.) *The Agrarian History of England and Wales, Vol. IV, 1500–1640*, (Cambridge: Cambridge University Press), pp. 113–60.

Emigh, R.J. (1997) 'The Spread of Sharecropping in Tuscany: the Political Economy of Transaction Costs', *American Sociological Review*, 62, 423–42.

Evans, E. (1976) *The Contentious Tithe: The Tithe Problem and English Agriculture, 1750–1850*, (Routledge and Kegan Paul).

Federico, G. (2006) 'The "Real" Puzzle of Sharecropping: Why is it Disappearing?', *Continuity and Change*, 21, 261–85.

Fisher, J.R. (2000) 'Agrarian Politics', in Collins, E.J.T. (ed.) *The Agrarian History of England and Wales, Vol. VII, 1850–1914, part I*, (Cambridge: Cambridge University Press), pp. 321–57.

Foster, C.F. (2002) *Seven Households: Life in Cheshire and Lancashire, 1582–1774*, (Northwich: Arley Hall Press).

Freeman, M. (2002) *Social Investigation and Rural England, 1870–1914*, (Woodbridge: Boydell Press).

Fussell, G.E. (ed.) (1936) *Robert Loder's Farm Accounts 1610–1620*, Camden Society, 3rd series, Vol. 53.

Galassi, F.L. (2000) 'Moral Hazard and Asset Specificity in the Renaissance: The Economics of Sharecropping in 1427 Florence', in Kauffman, K.D. (ed.) *Advances in Agricultural Economic History, Vol. 1, New Frontiers in Agricultural History*, (Stamford, CT: JAI Press), pp. 177–206.

Garnett, R.G. (1972) *Co-operation and the Owenite Socialist Communities in Britain, 1825–45*, (Manchester: Manchester University Press).

Gemmill, W. (2005) 'The Role of the New Entrant', *Journal of the Royal Agricultural Society of England*, 165.

George, H. (1880) *Progress and Poverty*, (Reeves).

Gibbard, R., Ravenscroft, N., and Reeves, J. (1999) 'The Popular Culture of Agri-Cultural Law Reform', *Journal of Rural Studies* 15, 269–78.

Gibberd, R. (1995) 'Farm Tenancies: Paradise Leased or Paradise Lost', *Journal of the Royal Agricultural Society of England*, 156, 106–13.

Goubert, P. (1969) *The Ancien Régime: French Society 1600–1750*, (Weidenfeld and Nicolson).

Gould, P. (1988) *Early Green Politics, 1880–1900*, (Brighton: Harvester Press).

Grey, A. (1891) 'Profit Sharing in Agriculture', *Journal of the Royal Agricultural Society of England*, 3rd series, 2, pp. 779–80.

Griffiths, E.M. (1987) 'The Management of Two East Norfolk Estates in the Seventeenth Century: Blickling and Felbrigg', (Unpublished PhD thesis, University of East Anglia, 1987).

Griffiths, E.M. (1988) 'Sir Henry Hobart: A New Hero of Norfolk Agriculture?', *Agricultural History Review*, 46, 15–34.

Griffiths, E.M. (2004) 'Responses to Adversity: The Changing Strategies of Two Norfolk Landowning Families, c. 1665–1700' in Hoyle, R.W. (ed.) *People, Landscape and Alternative Agriculture: Essays for Joan Thirsk*, Agricultural History Review Supplement 3 (2004), pp. 74–92.

Griffiths, E.M. (ed.) (2002) *William Windham's Green Book 1673–1688*, Norfolk Record Society, vol. LXVI.

Habakkuk, J. (1994) *Marriage, Debt, and the Estates System: English Landownership 1650–1950*, (Oxford: Clarendon Press).

Hadfield, A.M. (1970) *The Chartist Land Company*, (Newton Abbot: David & Charles).

Hall, A.D. (1913) *A Pilgrimage of British Farming, 1910–1912*, (Murray).

Hall, F.J. and Martyn, P. (1993) *Changes in Sharemilking: 1973–1993*, Technical Paper, 11 (Ministry of Agriculture, New Zealand).

Handley, J.E. (1953) *Scottish Farming in the Eighteenth Century*, (Faber).

Harrison, C.J. (1979) 'Elizabethan Village Surveys: A Comment', *Agricultural History Review*, 27, 82–9.

Harrison, G.V. (1984) 'The South West: Dorset, Somerset, Devon and Cornwall', in Thirsk, J. (ed.) *The Agrarian History of England and Wales, Vol. V, 1640–1750. I. Regional Farming Systems*, (Cambridge: Cambridge University Press), pp. 358–92.

Hartlib, S. (1652) *His Legacie or An Enlargement of the discourse of Husbandry Used in Brabant and Flanders*.

Harvey, B. (1969) 'The Leasing of the Abbot of Westminster's Demesne in the Later Middle Ages', *Economic History Review*, 2nd series, 22, 17–27.

Harvey, B. (1977) *Westminster Abbey and its Estates in the Middle Ages*, (Oxford: Clarendon Press).

Hatcher, J. (1970) *Rural Economy and Society in the Duchy of Cornwall, 1300–1500*, (Cambridge: Cambridge University Press).

Havinden, M.A. (1966) *Estate Villages: A Study of the Berkshire Villages of Ardington and Lockinge*, (Reading: Lund Humphries).

Henderson, G. (1943) *The Farming Ladder*, (Faber & Faber).

Higgs, H. (1894) 'Métayage in Western France', *The Economic Journal*, 4, 1–13.

Hill, B. and Gasson, R. (1985) 'Farm tenure and farming practice', *Journal of Agricultural Economics*, 36, 187–99.

Hilton, R.H. (1973) *The English Peasantry in the Later Middle Ages*, (Oxford: Clarendon Press).

Hilton, R.H. (1990) 'Why Was There so Little Champart Rent in Medieval England?', *Journal of Peasant Studies*, 17, 509–19.

Hobsbawm, E.J. and Rudé, G. (1969) *Captain Swing*, (Lawrence & Wishart).

Hoffman, P. (1982) 'Sharecropping and Investment in Agriculture in Early Modern France', *Journal of Economic History*, 42, 155–9.

Hoffman, P. (1984) 'The Economic Theory of Sharecropping in Early Modern France', *Journal of Economic History*, 309–19.

Hoffman, P. (1996) *Growth in a Traditional Society: the French Countryside 1450–1815*, (Princeton: Princeton University Press).

Holderness, B.A. (1972) 'The Agricultural Activities of the Massingberds of South Ormsby, Lincolnshire 1638–c.1750', *Midland History*, 1, 15–25.

Holderness, B.A. (1976) 'Credit in English Rural Society before the Nineteenth Century with Special Reference to the Period 1650–1720', *Agricultural History Review*, 24, 97–109.

Holderness, B.A. (1984) 'East Anglia and the Fens: Norfolk, Suffolk, Cambridgeshire, Ely, Huntingdonshire, Essex and the Lincolnshire Fens', in Thirsk, J. (ed.) *The Agrarian History of England and Wales, Vol. V, 1640–1750. I. Regional Farming Systems*, (Cambridge: Cambridge University Press), pp. 197–238.

Homans, G.C. (1942) *English Villagers of the Thirteenth Century*, (Cambridge, Mass.: Harvard University Press).

Horn, C.A. (1975) 'Two Centuries of Incentive Payments in Agriculture', *The Accountants' Review*, 26, 223–9.

Horn, P. (1978) 'The Dorset Dairy System', *Agricultural History Review*, 26, 100–7.

Howell, C. (1983) *Land, Family and Inheritance in Transition: Kibworth Harcourt*, (Cambridge: Cambridge University Press).

Howkins, A. (2000) 'Social, Cultural and Domestic Life', in Collins, E.J.T. (ed.) *The Agrarian History of England and Wales, Vol. VII, 1850–1914, part II*, (Cambridge: Cambridge University Press), pp. 1354–424.

Howkins, A. (2000) 'Types of Rural Communities', in Collins, E.J.T. (ed.) *The Agrarian History of England and Wales, Vol. VII, 1850–1914, part II*, (Cambridge: Cambridge University Press), pp. 1297–353.

Hoyle, R.W. (2001) 'Agrarian Agitation in Mid-Sixteenth-Century Norfolk: A Petition of 1553', *Historical Journal*, 44, pp. 223–38.

Hunt, E.H. and Pam, S.J. (2003) 'Responding to Agricultural Depression, 1873–96, Managerial Success, Entrepreneurial Failure', *Agricultural History Review*, 50, 225–52.

Hunt, J. (2005) 'Landowner Profile', *CLA Magazine*, (Country Landowners Association).

James, J.F. and Bettey, J.H. (eds) (1993) *Farming in Dorset: Diary of James Warne, 1758; letters of George Boswell, 1787–1805*, Dorset Record Society, vol. 13.

Jenkins, P. (1983) *The Making of a Ruling Class: The Glamorgan Gentry, 1640–1790*, (Cambridge: Cambridge University Press).

Johnson, R.W.M. (2000) *Reforming EU Farm Policy: Lessons from New Zealand*, (Institute of Economic Affairs).

Keith, J. (1954) *Fifty Years of Farming* (Faber & Faber).

Kerr, B. (1968) *Bound to the Soil*, (John Baker).

Kerridge, E. (1967) *The Agricultural Revolution*, (George Allen & Unwin).

Kerridge, E. (1969) *Agrarian Problems in the Sixteenth Century and After*, (George Allen & Unwin).

Kerridge, E. (1973) *The Farmers of Old England*, (George Allen & Unwin).

Kettle, A.J. (1989) 'Agriculture 1300–1540', in Baugh, G.C. (ed.) *The Victoria History of the County of Shropshire, Vol. 4, Agriculture*, (Oxford: Oxford University Press for the Institute of Historical Research), pp. 72–118.

Ketton-Cremer, R.W. (1962) *Felbrigg: The Story of a House*, (Ipswich: The Boydell Press).

Ketton-Cremer, R.W. (1969) *Norfolk in the Civil War*, (Faber & Faber).

Lennard, R.V. (1959) *Rural England 1086–1135: A Study of Social and Agrarian Conditions*, (Oxford: Clarendon Press).

Lewis, C.P. (2002) 'Woodditton', in Wareham, A.F. and Wright, A.P.M. (eds) *A History of the County of Cambridge and the Isle of Ely, Vol. X, North-eastern Cambridgeshire*, (Oxford: Oxford University Press for the Institute of Historical Research).

Little, M. (2005) 'Joint and Several: Dealing with Farm Agreements Post CAP Reform', *Savills Aspects of Land East*.

MacAskill, J. (1959) 'The Chartist Land Plan', in A. Briggs (ed.) *Chartist Studies*, (Macmillan), pp. 304–41.

Marsden, T. (1986) 'Property-state Relations in the 1980s: An Examination of Landlord-Tenant Legislation in British Agriculture', in Cox, G., Lowe, P. and Winter, M. (eds) *Agriculture: People and Policies*, (Allen and Unwin), pp. 126–45.

Marsh, J. (1892) *Back to the Land: The Pastoral Impulse in England from 1880–1914*, (Quartet Books).

Marshall, A. (1961) *Principles of Political Economy*, 8th edn, (Macmillan).

Marshall, W. (1787) *The Rural Economy of Norfolk*, 2 vols.

Marshall, W. (1796) *The Rural Economy of the West of England*, 2 vols (1796).

Massingberd, W.O. (1893) *History of the Parish of Ormsby-cum-Ketsby*, (Lincoln).

McIntosh, M.K. (1980) 'Land, Tenure and Population in the Royal Manor of Havering, Essex, 1251–1352', *Economic History Review*, 2nd series, 33, 17–31.

McQuiston, J.R. (1973) 'Tenant Right: Farmer against Landlord in Victorian England, 1847–1883', *Agricultural History*, 47, 95–113.

Mill, J.S. (1848) *Principles of Political Economy*.

Miller, E. (1991) 'Tenant Farming and Tenant Farmers: The Southern Counties', in Miller, E. (ed.) *The Agrarian History of England and Wales, Vol. III, 1348–1500*, (Cambridge: Cambridge University Press), pp. 703–22.

Miller, E. and Hatcher, J. (1978) *Medieval England: Rural Society and Economic Change, 1086–1348*, (Longman).

Mills, D.R. (1980) *Lord and Peasant in Nineteenth-Century Britain*, (Croom Helm).

Mingay, G.E. (1956) 'The Agricultural Depression, 1730–1750', *Economic History Review*, 2nd series, 8, 323–38.

Mingay, G.E. (1962) 'The Size of Farms in the Eighteenth Century', *Economic History Review*, 2nd series, 14 (1962) 469–88.

Mingay, G.E. (1990) 'The Diary of James Warne, 1758', *Agricultural History Review*, 38, 72–8.

Mitchison, R. (1959) 'The Old Board of Agriculture, 1793–1822', *English Historical Review*, 74, 41–69.

Moreton, C.E. (1992) *The Townshends and their World: Gentry, Law, and Land in Norfolk c.1450–1551*, (Oxford: Clarendon Press).

Moriceau, J.M. (1994) 'Fermayage and Métayage (XII–XIX siècle)', *Histoire et Sociétiés Rurales*, 1, 155–90.

Murdoch, J. and Ward, N. (1997) 'Governmentality and territoriality: The Statistical Manufacture of Britain's "National Farm"', *Political Geography*, 16 (1997) 307–24.

Newbery, D.M.G. (1977) 'Risk Sharing, Sharecropping and Uncertain Labour Markets', *Review of Economic Studies*, 2nd series, 44, 585–94.

Newby, H., Bell, C., Rose, D. and Saunders, P. (1978) *Property Paternalism and Power*, (Hutchinson).

Nix, J., Hill, P. and Williams, N. (1987) *Land and Estate Management*, (Chichester: Packard).

Ó Gráda, C. (1974) 'The Owenite Community at Ralahine, County Clare, 1831–1833: A Reassessment', *Irish Economic and Social History*, 1, 36–48.

O'Neill, K. (1984) *Family and Farm in Pre-Famine Ireland: the Parish of Killashandra*, (University of Wisconsin Press).

Oestmann, C. (1994) *Lordship and Community: the Lestrange Family and the Village of Hunstanton in the First Half of the Sixteenth Century*, (Woodbridge: The Boydell Press).

Offer, A. (1991) 'Farm Tenure and Land Values in England, c.1750–1950', *Economic History Review*, 2nd series, 44, 1–20.

Ogilvie, S. (2007) '"Whatever is, is Right"? Economic Institutions in Pre-Industrial Europe', *Economic History Review*, 2nd series, 60, 649–84.

Overton, M. (1981) 'Agricultural Change in Norfolk and Suffolk, 1580–1740', (Unpublished PhD thesis, University of Cambridge).

Overton, M. (1986) 'Agriculture', in Langton, J. and Morris, R.J. (eds) *Atlas of Industrializing Britain*, (Methuen), pp. 34–53.

Overton, M. (1996) *Agricultural Revolution in England: The Transformation of the Agrarian Economy, 1500–1800*, (Cambridge: Cambridge University Press).

Overton, M. and Campbell, B.M.S. (1992) 'Norfolk Livestock Farming 1250–1740: A Comparative Study of Manorial Accounts and Probate Inventories', *Journal of Historical Geography*, 18, 377–96.

Overton, M., Whittle, J., Dean, D. and Hann, A. (2004) *Production and Consumption in English Households, 1600–1750*, (Routledge).

Papworth, A.J. (2003) 'Diversity and Diversification', *Journal of the Royal Agricultural Society of England*, 164. [http://www.rase.org.uk/activities/publications/RASE_journal/2003/09-58151622.pdf].

Peacock, A.J. (1965) *Bread or Blood: A Study of the Agrarian Riots in East Anglia in 1816*, (Gollancz).

Perkins, J.A. (1975) 'Tenure, Tenant Right and Agricultural Progress in Lindsey, 1780–1850', *Agricultural History Review*, 23, 1–22.

Perren, R. (1989) 'Agriculture 1875–1985', in G.C. Baugh (ed.) *The Victoria History of the County of Shropshire, Vol. 4, Agriculture*, (Oxford: Oxford University Press for the Institute of Historical Research), pp. 232–69.

Perren, R. (1995) *Agriculture in Depression 1870–1940*, (Cambridge: Cambridge University Press).

Pickard, J. (2008) 'Shepherding in Colonial Australia', *Rural History*, 19, 55–80.

Plumb, J.H. (1956) *Sir Robert Walpole. [1], The Making of a Statesman*, (Cressett Press).

Postan, M.M. (1954) 'Introduction', in King, P.I. (ed.) *The Book of William Morton: Almoner of Peterborough Monastery, 1448–1467*, Northants Records Society, 16, pp. xi–xlviii.

Postan, M.M. (1972) *The Medieval Economy and Society*, (Weidenfeld and Nicolson).

Postan, M.M. (1973) *Essays on Medieval Agriculture and General Problems of the Medieval Economy*, (Cambridge: Cambridge University Press).

Prothero, R.E. (1908) *The Pleasant Land of France*.

Raftis, J.A. (1964) *Tenure and Mobility: Studies in the Social History of the Mediaeval English Village*, (Toronto: Pontifical Institute of Mediaeval Studies).

Ravensdale, J. (1984) 'Population Changes and the Transfer of Customary Land on a Cambridgeshire Manor [Cottenham] in the Fourteenth Century', in Smith, R.M. (ed.) *Land, Kinship and Life Cycle*, (Cambridge: Cambridge University Press), pp. 197–225.

Razi, Z. (1981) 'Family, Land and the Village Community in Later Medieval England', *Past & Present*, 93, 3–36.

Reed, M. (1984) 'The Peasantry of Nineteenth Century England: A Neglected Class?', *History Workshop Journal*, 18, 53–76.

Richmond, C. (1981) *John Hopton: A Fifteenth Century Suffolk Gentleman*, (Cambridge, Cambridge University Press).

Richmond, C. (1996) 'Landlord and Tenant: The Paston Evidence', in Kermode, J. (ed.) *Enterprise and Individuals in Fifteenth Century England*, (Stroud: Alan Sutton), pp. 25–42.

Rose, D., Newby, H., Saunders, P. and Bell, C. (1977) 'Land Tenure and Official Statistics: A Research Note', *Journal of Agricultural Economics*, 28, 69–75.

Rosenheim, J.M. (1989) 'Count Governance and Elite Withdrawal in Norfolk, 1660–1720', in Beier, A.L., Cannadine, D. and Rosenheim, J.M. (eds), *The First*

Modern Society: Essays in English History in Honour of Lawrence Stone, (Cambridge: Cambridge University Press), pp. 106–15.

Rosenheim, J.M. (1989) *The Townshends of Raynham: Nobility in Transition in Restoration and Early Hanoverian England*, (Middletown, Ct.: Wesleyan University Press).

Rosenheim, J.M. (1998) *The Emergence of a Ruling Order: English Landed Society 1650–1750*, (Longman).

Royal Commission on Labour Reports by C.M. Chapman, Berkshire, BPP, 1893–4, XXXV, c.6894–II.

Royal Commission on Labour, Reports by A. Wilson Fox, Northumberland, BPP, 1893–4, XXXV, c.6894–III.

Santos, R. (2006) 'Risk-sharing and Social Differentiation of Demand in land Tenancy Markets in Southern Portugal, seventeenth–Nineteenth Centuries', *Continuity and Change*, 21, 287–312.

Schofield, P. and Câmara, B. (2006) 'Introduction, Sharecropping in History', *Continuity and Change*, 21, 209–11.

Shaw-Lefevre, G. (1893) *Agrarian Tenures: A Survey of the Laws and Customs Relating to the Holding of Land in England, Ireland and Scotland*, (Cassell).

Shaw-Taylor, L. (2001) 'Labourers, Cows, Common Rights and Parliamentary Enclosure: The Evidence of Contemporary Comment c. 1760–1810', *Past & Present*, 171, 95–126.

Shaw-Taylor, L. (2001) 'Parliamentary Enclosure and the Emergence of an English Agricultural Proletariat', *Journal of Economic History*, 61, 640–62.

Short, B., Watkins, C., Foot, W. and Kinsman, P. (2000) *The National Farm Survey 1941–1943: State Surveillance and the Countryside in England and Wales in the Second World War*, (Wallingford: CABI).

Simpson, A. (1958) 'The East Anglian Foldcourse: Some Queries', *Agricultural History Review*, 6, 87–96.

Simpson, A. (1963) *The Wealth of the Gentry 1540–1660: East Anglian Studies*, (Cambridge: Cambridge University Press).

Sinclair, K. (2000) *A History of New Zealand* (Penguin).

Smith, A. (1776) *An Inquiry into the Nature and Causes of the Wealth of Nations*, Book 3, Part 2.

Smith, A.H., Baker, G.M. and Kenny, R.W. (eds) (1979) *The Papers of Nathaniel Bacon of Stiffkey, vol. 1, 1556–1577*, Norfolk Record Society, vol. 46.

Smith, A.H., Baker, G.M. and Kenny, R.W. (eds) (1983) *The Papers of Nathaniel Bacon of Stiffkey, Vol. 2, 1578–1585*, Norfolk Record Society, vol. 49.

Smith, R.M. (1984) 'Families and Their Land in an Area of Partible Inheritance: Redgrave, Suffolk, 1260–1320', in Smith, R.M. (ed.) *Land, Kinship and Life Cycle*, (Cambridge: Cambridge University Press), pp. 135–95.

Smith, R.M. (1984) 'Some Issues Concerning Families and Their Property in Rural England, 1250–1800', in Smith, R.M. (ed.) *Land, Kinship and Life Cycle*, (Cambridge: Cambridge University Press), pp. 1–86.

Smout, T.C. (1969) *A History of the Scottish People, 1560–1830*, (Collins).

Snowden, F.M. (1989) *The Fascist Revolution in Tuscany, 1919–1922*, (Cambridge: Cambridge University Press).

Speed, A. (1659) *Adam out of Eden*.

Spufford, M. (1974) *Contrasting Communities: English Villagers in the Sixteenth and Seventeenth Centuries*, (Cambridge: Cambridge University Press).

Stanes, R. (1990) *The Old Farm: A History Of Farming Life in the West Country*, (Exeter: Devon Books).

Stead, D.R. (2004) 'Risk and Risk Management in English Agriculture, c.1750–1850', *Economic History Review*, 2nd series, 57, 334–61.

Steele, E.D. (1974) *Irish Land and British Politics: Tenant Right and Nationality, 1865–1870*, (Cambridge: Cambridge University Press).

Stevenson, W. (1815) *General View of the Agriculture of the County of Dorset*.

Stone, D. (2003) 'Productivity and Sheepfarming in late Medieval England', *Agricultural History Review*, 51, 1–22.

Stratton, R. (1983) *Sharefarming*, (Country Landowners Association).

Stratton, R., Sydenham, A., and Bird, A. (1992) *Sharefarming: The Practice – With Model Form of Agreement*, 3rd edn (Country Landowners Association).

Study of Joint Venture Farming. Report to DEFRA (2007) (Wolverhampton: ADAS).

Sturmey, S.G. (1955) 'Owner-farming in England and Wales 1900–1950', *Manchester School*, 23, 246–68.

Suggett, R. (2005) *Houses and History on the March: Radnorshire 1400–1800*, (Ceredigion: RCAHMW).

Taylor, G. (1996) *A Review of Sharemilking: 1972–1996*, (Rural Policy Unit, Ministry of Agriculture, New Zealand).

Taylor, S. (1884) *Profit-sharing between Capital and Labour*, (Kegan Paul).

Tenanted Land Survey – England 2006, DEFRA stats 09/07 [https://statistics.defra.gov.uk/esg/statnot/astl.pdf] accessed 11 August 2008.

Thirsk, J. (1967) 'The Farming Regions of England', in *idem* (ed.) *The Agrarian History of England and Wales, Vol. IV, 1500–1640*, (Cambridge: Cambridge University Press), pp. 1–112.

Thirsk, J. (1997) *Alternative Agriculture: A History from the Black Death to the Present Day* (Oxford: Clarendon Press).

Thirsk, J. (ed.) (1967–2000) *The Agrarian History of England and Wales*, 8 vols, (Cambridge: Cambridge University Press).

Thorold Rogers, J.E. (1912) *Six Centuries of Work and Wages: The History of English Labour*, (11th edn, T. Fisher Unwin).

Townsend, R.M. (1993) *The Medieval Village Economy*, (Princeton: Princeton University Press).

Troup, D.A.G. (1984) *Agricultural Holdings Act, 1984: The Practitioner's Companion*, (Surveyor's Publication).

Turgot, A.R.J. (1766) *Reflections on the Formation and Distribution of Wealth*, (Paris).

Turner, M. (1986) 'The Land Tax, Land, and Property: Old Debates and New Horizons', in Turner, M. and Mills, D. (eds) *Land Tax and Property: The English Land Tax, 1692–1832*, (Gloucester: Alan Sutton), pp. 1–35.

Turner, M.E., Beckett, J.V., and Afton, B. (1997) *Agricultural Rent in England, 1690–1914*, (Cambridge: Cambridge University Press).

Venn, J.A. (1923) *The Foundations of Agricultural Economics*, (Cambridge: Cambridge University Press).

Wade Martins, S. and Williamson, T. (1998) 'The Development of the Lease and its Role in Agricultural Improvement in East Anglia, 1660–1870', *Agricultural History Review*, 46, 129–30.

Wallace, A.R. (1882) *Land Nationalization: Its Necessity and Aims*, (Trubner).

Walton, R. and Essayon, M. (1982) *Adkins's Landlord and Tenant*, 18th edn (The Estates Gazette Ltd).

Whitehead, I., Errington, A., Millard, N. and Felton, T. (2002) *An Economic Evaluation of the Agricultural Tenancies Act 1995*, (University of Plymouth Report to DEFRA).

Whittle, J. (2000) *The Development of Agrarian Capitalism: Land and Labour in Norfolk 1440–1580*, (Oxford: Clarendon Press).

Wilson, A.R. (1995) *Forgotten Harvest: The Story of Cheesemaking in Wiltshire*, (Calne: the author).

Winter, M. (2007) 'Revisiting Landownership and Property Rights', in Clout, H. (ed.) *Contemporary Rural Geographies, Land, Property and Resources in Britain: Essay in Honour of Richard Munton*, (University College London Press), pp. 72–83.

Winter, M., Richardson, C., Short, C. and Watkins, C. (1990) *Agricultural Land Tenure in England and Wales*, (Royal Institution of Chartered Surveyors).

Woodward, D. (ed.) (1984) *The Farming and Memoranda Books of Henry Best of Elmswell, 1642*, Records of Social and Economic History, new series, 8.

Worgan, G.B. (1811) *General View of the Agriculture of the County of Cornwall*.

Wrightson, K. (2000) *Earthly Necessities: Economic Lives in Early Modern Britain*, (Yale University Press).

Young, A. (1770) *A Farmer's Guide to Hiring and Stocking Farms*.

Young, A. (1804) *General View of the Agriculture of the County of Norfolk*.

Young, A. (1808) *General Report on Enclosures*.

Young, A. (1929) *Travels Through France During the Years 1787, 1788 and 1789*, ed. C. Maxwell, (Cambridge: Cambridge University Press).

Index